SPECIATION OF THE WANDERING SHREW

JAMES S. FINDLEY

ZINC READ

Published by
This edition copyright © 2024 by Zinc Read
Design and setting By Zinc Read
Email- zincread@gmail.com
ISBN-978-93-6277-404-0

CONTENTS

INTRODUCTION iv

MATERIALS METHODS AND ACKNOWLEDGMENTS 1

NON-GEOGRAPHIC VARIATION 5

PELAGE CHANGE 8

GEOGRAPHIC DISTRIBUTION AND VARIATION 9

ORIGIN OF THE *SOREX VAGRANS* RASSENKREIS 17

RELATIONSHIPS WITH OTHER SPECIES 28

CONCLUSIONS 74

INTRODUCTION

The purpose of this report is to make clear the biological relationships between the shrews of the *Sorex vagrans-obscurus* "species group." This group as defined by H. H. T. Jackson (1928:101) included the species *Sorex vagrans, S. obscurus, S. pacificus, S. yaquinae,* and *S. durangae.* The last mentioned species has been shown (Findley, 1955:617) to belong to another species group. *Sorex milleri,* also assigned to this group by Jackson (1947:131), seems to have its affinities with the *cinereus* group as will be explained beyond. The position of the *vagrans* group in relationship to other members of the genus will be discussed.

Of this group, the species that was named first was *Sorex vagrans* Baird, 1858. Subsequently many other names were based on members of the group and these names were excellently organized by Jackson in his 1928 revision of the genus. Subsequent students of western mammals, nevertheless, have been puzzled by such problems as the relationship of (1) *Sorex vagrans monticola* to *Sorex obscurus obscurus* in the Rocky Mountains, (2) *Sorex pacificus, S. yaquinae,* and *S. obscurus* to one another on the Pacific Coast, and (3) *S. o. obscurus* to *S. v. amoenus* in California. Few studies have been made of these relationships. Clothier (1950) studied *S. v. monticola* and *S. o. obscurus* in western Montana and concluded that the two supposed kinds actually were not separable in that area. Durrant (1952:33) was able to separate the two kinds in Utah as was Hall (1946:119, 122) in Nevada. Other mammalogists who worked within the range of the *vagrans-obscurus* groups have avoided the problems in one way or another. Recently Rudd (1953) has examined the relationships of *S. vagrans* to *S. ornatus.*

MATERIALS METHODS AND ACKNOWLEDGMENTS

Approximately 3,465 museum study skins and skulls were studied. Most of these were assembled at the University of Kansas Museum of Natural History, but some were examined in other institutions.

Specimens were grouped by geographic origin, age, and sex. Studies of the role of age and sex in variation were made. Because it was discovered that secondary sexual variation was negligible, both males and females, if of like age and pelage, were used in comparisons designed to reveal geographic variation.

External measurements used were total length, length of tail, and length of hind foot. After studying a number of cranial dimensions I chose those listed below as the most useful in showing differences in size and proportions of the skull. Figures 1 and 2 show the points between which those measurements were taken.

Condylobasal length.--From anteriormost projection of the premaxillae to posteriormost projection of the occipital condyles (a to a).

Maxillary tooth-row.--From posteriormost extension of M3 to anteriormost extension of first unicuspid (b to b).

Palatal length.--From anteriormost projection of premaxillae to posteriormost part of bony palate (c to c).

Cranial breadth.--Greatest lateral diameter of braincase (d to d).

Least interorbital breadth.--Distance between medialmost superior edges of orbital fossae, measured between points immediately above and behind posterior openings of infraorbital foramina (e to e).

Maxillary breadth.--Distance between lateral tips of maxillary processes (f to f).

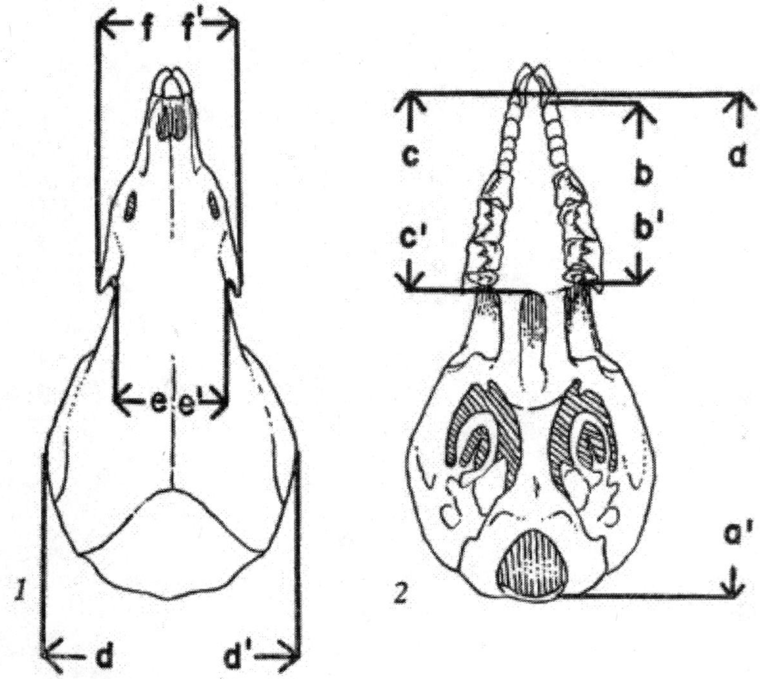

FIGS. 1 AND 2. Showing where certain cranial measurements were taken. x 31/2. (Based on *Sorex vagrans obscurus*, from Stonehouse Creek, 51/2 mi., W junction of Stonehouse Creek and Kelsall River, British Columbia, , 28545 KU.)

In descriptions of color, capitalized terms refer to those in Ridgway (1912). In addition the numerical and alphabetical designations of these terms are given since a knowledge of the arrangements of these designations enables one quickly to evaluate differences between stated colors. Color terms which are not capitalized do not refer to any precise standard of color nomenclature.

In the accounts of subspecies, descriptions, unless otherwise noted, are of first year animals as herein defined. Descriptions of color are based on fresh pelages.

Unless otherwise indicated, specimens are in the University of Kansas Museum of Natural History. Those in other collections are identified by the following abbreviations:

AMNH

American Museum of Natural History

CM

Carnegie Museum
ChM
Chicago Museum of Natural History
CMNH
Cleveland Museum of Natural History
FC
Collection of James S. Findley
HC
Collection of Robert Holdenreid
SGJ
Collection of Stanley G. Jewett
CDS
Collection of Charles D. Snow
AW
Collection of Alex Walker
NMC
National Museum of Canada
OSC
Oregon State College
PMBC
British Columbia Provincial Museum of Natural History
SD
San Diego Natural History Museum
BS
United States Biological Surveys Collection
USNM
United States National Museum
UM
University of Michigan Museum of Zoology
OU
University of Oregon Museum of Natural History
UU
University of Utah Museum of Zoology
WSC
Washington State College, Charles R. Conner Museum

In nature, the subspecies of *Sorex vagrans* form a cline and are distributed geographically in a chain which is bent back upon itself. The subspecies in the following accounts are listed in order from the southwestern end of the chain clockwise back to the zone of overlap.

The synonymy of each subspecies includes the earliest available name and other names in chronological order. These include the first usage of the name combination employed by me and other name combinations that have been applied to the subspecies concerned.

In the lists of specimens examined, localities are arranged first by state or province. These are listed in tiers from north to south and in any given tier from west to east. Within a given state, localities are grouped by counties, which are listed in the same geographic sequence as were the states and provinces (N to S and W to E). Within a given county, localities are arranged from north to south. If two or more localities are at the same latitude the westernmost is listed first. Marginal localities are listed in a separate paragraph at the end of each account. The northernmost marginal locality is listed first and the rest follow in clockwise order. Those records followed by a citation to an authority are of specimens which I have not personally examined. Marginal records are shown by dots on the range maps. Marginal records which cannot be shown on the maps because of undue crowding are listed in Italic type.

To persons in charge of the collections listed above I am deeply indebted. Without their generous cooperation in allowing me to examine specimens in their care this study would not have been possible. Appreciated suggestions in the course of the work have been received from Professors Rollin H. Baker, A. Byron Leonard, R. C. Moore, Robert W. Wilson, and H. B. Tordoff, and many of my fellow students. Mr. Victor Hogg gave helpful suggestions on the preparation of the illustrations. My wife, Muriel Findley, devoted many hours to secretarial work and typing of manuscript. Finally I am grateful to Professor E. Raymond Hall for guidance in the study and for assistance in preparing the manuscript. During the course of the study I received support from the University of Kansas Endowment Association, from the Office of Naval Research, and from the National Science Foundation.

NON-GEOGRAPHIC VARIATION

Non-geographic variation, that is to say, variation within a single population of shrews, consists of variation owing to age and normal individual variation. In *Sorex* I have detected no significant secondary sexual differences between males and females; accordingly the two sexes are here considered together.

Variation with age must be considered in order to assemble comparable samples of these shrews. Increased age results in wear on all teeth and in particularly striking changes in the size and shape of the first incisors. Skulls of older shrews develop sagittal and lambdoidal ridges, and further differ from skulls of young animals in being slightly broader and shorter, and in developing thicker bone, particularly on the rostrum which thus seems to be, but is not always in fact, more robust. Pruitt has recently (1954) noted these same cranial differences in specimens of *Sorex cinereus* of different ages.

Several students of American shrews, notably Pearson (1945) on *Blarina*, Hamilton (1940) on *Sorex fumeus*, and Conaway (1952) on *Sorex palustris*, have shown that young are born in spring and summer, usually reach sexual maturity the following spring, and rarely survive through, or even to, a second winter. The result is that collections made, as most of them are, in spring and summer, contain two age classes, first year and second year animals. These two age classes are readily separable on the basis of differences in the skull as well as on the decreased pubescence of the tail and the increased weight of second year animals. My own examination of hundreds of museum specimens confirms this for the *Sorex vagrans* group. Separation of the two age classes in an August-taken series of *Sorex vagrans* from coastal Washington is shown in figure 3, in which two tooth-measurements that are dependent upon wear are plotted against one another.

First year animals are more abundant in collections than are second year animals. Within the first year, that is to say from spring to late fall, animals vary but little. Dental characters are best studied in first year shews. For this reason I have used them as the basis for the study of geographic variation, and descriptions are based on first year animals unless otherwise noted.

CHARACTERS OF TAXONOMIC WORTH

Within the *Sorex vagrans* complex, the only characters of taxonomic significance that I have detected are in size and color. It is true that cranial proportions, such as relative size of rostrum, may change from population to population, but these proportions seem to me to be dependent upon actual size of the individual shrew as I shall elsewhere point out. Of the cranial measurements here employed, palatal length and least interorbital breadth are the most significant and useful. Color in the *S. vagrans* group seems to be in Orange and Cadmium Yellow, colors 15 and 17 of Ridgway (1912). No specimens actually possess these pure colors, but most colors in these shrews are seen to be derived from the two mentioned by admixture of black and/or neutral gray. In color designations an increase in neutral gray is indicated by an increased number of prime signs ( ), whereas increase in black is indicated by progressive characters of the Roman alphabet (*i, k, m*). Thus, 17 *k* is grayer than 17 *k* and 17 *m* is blacker than 17 *k*. In subspecific diagnoses in this report, color and size, and sometimes relative size, are the characters usually mentioned.

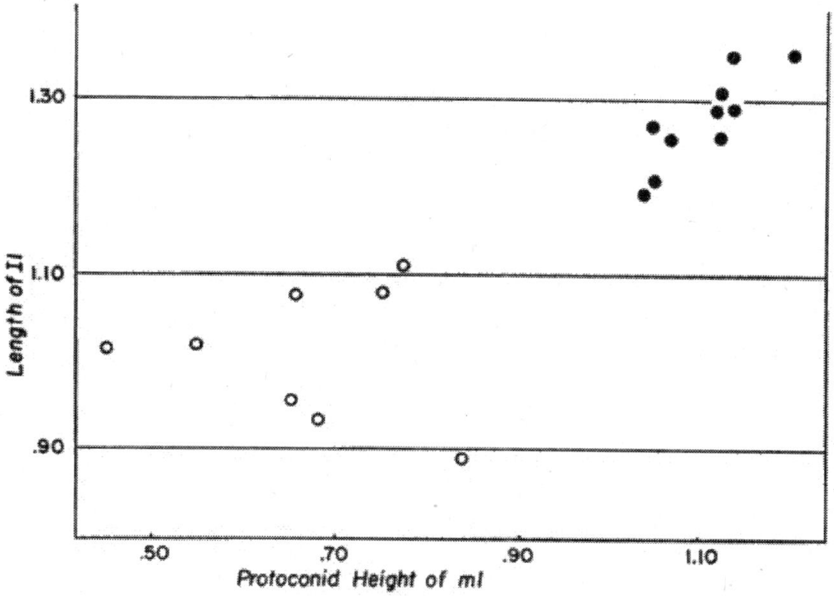

FIG. 3. Two measurements (in millimeters) reflecting tooth-wear plotted against one another. First year and second year individuals of *Sorex vagrans vagrans,* all taken in August at Willapa Bay, Washington, are completely separated. Open circles represent teeth of second year shrews; solid circles represent teeth of first year shrews.

PELAGE CHANGE

In general, winter pelage is darker than summer pelage in these shrews. Winter pelage comes in first on the rump and spreads caudad and ventrad. The growth line of incoming hair is easily detected on the fur side of the skin. Throughout the winter the color of the pelage changes, often becoming somewhat browner, although no actual molt takes place. This was noted by Dalquest (1944) who assumed that the color change resulted from molt although he was unable to detect actual replacement of hairs. Summer pelage usually comes in first on the back or head and moves posteriorly and laterally. Time of molt depends on latitude and altitude. Summer pelage may appear fairly late in the season and may account for the anomalous midsummer molt noted by Dalquest. Fresh pelages of summer and winter are best seen in first year animals and are less variable than are worn pelages and hence are used as the basis of color descriptions.

GEOGRAPHIC DISTRIBUTION AND VARIATION

Pacific Coastal Section

The largest shrews of the *vagrans* group (large in all dimensions) occur in the coastal forests of northern California and of Oregon. Those shrews are reddish, large-skulled, large-toothed, and have rostra that are large in proportion to the size of the skull as a whole. The very largest of these shrews live along the coast of northwestern California. To the southward they are somewhat smaller, and at successively more northern localities, to as far as southwestern British Columbia, they are likewise progressively smaller and also somewhat less reddish. The relative size of the rostrum decreases with the decrease in size of the skull; consequently smaller shrews have relatively smaller rostra (see fig. 4). In addition the zygomatic ridge of the squamosal decreases in relative size with decrease in actual size of the skull. Thus, these features change in a clinal fashion as one proceeds from, say, Humboldt County, California, northward to Astoria, Oregon.

Turning our attention now farther inland to the Cascade Mountains of northern Oregon, the shrews there also are smaller and less reddish (more brownish) than in northwestern California, and the trend to smaller and darker shrews culminates in the northern Cascades of Washington. Shrews from there, and from the southwestern coast of British Columbia, compared with those from northwestern California, are much smaller and have so great a suffusion of black that they appear brown rather than red. At places along the coast successively farther north of southwestern British Columbia the shrews become larger again, the largest individuals being those from near Wrangell, Alaska. From that place northwesterly along the coast of Alaska, size decreases again.

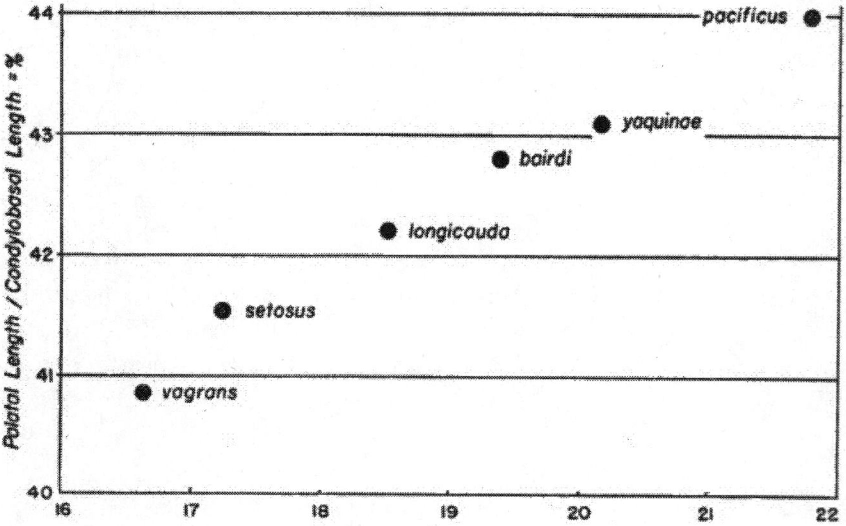

FIG. 4. Condylobasal length (in millimeters) plotted against palatal index (palatal length/condylobasal length x 100) in several subspecies of *Sorex vagrans* to show relative increase in size of rostrum with actual increase in size of skull.

The shrews so far discussed inhabit forests in a region of high rainfall and a minimum of seasonal fluctuation in temperature. Such a habitat seems to be the optimum for shrews of the *vagrans* group since the largest individuals are found there. In addition, shrews seem to be as common, or commoner, in this coastal belt, than they are in other places.

The large shrews of the *vagrans* group on the Pacific coast were divided into three species by H. H. T. Jackson in his revision of the North American *Sorex* in 1928. The large reddish shrews of the coast of California and southern Oregon were called *S. pacificus*. The somewhat smaller ones from the coast of central Oregon were called *S. yaquinae*. Still smaller shrews from northwestern Oregon and from the rest of the Pacific coast north into Alaska were called *S. obscurus*. I find these kinds to intergrade continuously one with the next in the manner described and conclude that all are of a single species.

Inland Montane Section

Inland from the coasts of British Columbia and Alaska the size of the *vagrans* shrew decreases rapidly. Specimens from western Alaska, central Alaska, and the interior of British Columbia are uniformly smaller than coastal specimens. In addition the red of the hair is masked more by neutral gray than by black with the result that the pelage is grayish rather than brownish or reddish. Shrews of this general appearance are found southward through the Rocky Mountain chain to Colorado and New Mexico. On the more or less isolated mountain ranges of Montana east of the continental divide the *vagrans* shrew is somewhat smaller still. On the Sacramento Mountains of southeastern New Mexico the shrew is somewhat larger and slightly darker. Southwestward from the Colorado Rockies this shrew becomes smaller and slightly more reddish (less grayish).

All of these montane populations of the *vagrans* shrew are commonest in hydrosere communities, that is to say, streamsides and marshy areas where the predominant vegetation is grass, sedges, willows, and alders. Since these animals are less common within the montane forests, hydrosere communities, rather than the actual forest, seem to be the positive feature important for the shrews.

The shrews of the montane region just described were regarded by Jackson as belonging to two species: *Sorex obscurus*, occupying all the Rocky Mountains south to, and including, the Sacramento Mountains; *S. vagrans*, made up of small individuals from various places in Wyoming, Montana, and Colorado, and all the shrews of western New Mexico and all of Arizona. My study of these animals has led me to conclude that the smaller shrews of Arizona and New Mexico intergrade in a clinal fashion with the shrews of Colorado and in fact represent but one species. Since some individuals from Colorado are as small as larger individuals from this southwestern population of small animals, I conclude that such specimens are the basis for reports of *S. vagrans* from Colorado. The shrews of the Sacramento Mountains resemble those of the Colorado Rockies more than they do the smaller shrews of western New Mexico and Arizona, possibly because the climate is similar in the Sacramento Mountains and the higher Colorado Rockies. There is less precipitation in the more western mountain ranges in New Mexico and in Arizona in April, May, and June than in the Colorado Rockies. These months are critical for the reproduction and growth of shrews.

As mentioned above, the shrews from east of the continental divide in Montana are smaller than those of the other mountains of the state, and it is upon such small animals that the name *Sorex vagrans* has been based in this area. It is clear, however, that these smaller animals intergrade with the larger shrews of the more western mountains. The small size might be an adaptation to the lesser precipitation and harsher continental climate east of the continental divide in Montana.

Great Basin and Columbia Plateau Section

The vagrant shrews of the Great Basin and adjoining Columbia Plateau and Snake River Plains are smaller than their relatives in the Rocky Mountains and, by virtue of less gray in their pelage, are reddish in summer and blackish rather than grayish in winter. There is little significant geographic variation in shrews throughout this region, although owing to their restriction to the vicinity of water, the populations of shrews are more or less isolated from one another and each is somewhat different from the next. Those from nearest the Rockies are sometimes slightly larger and those from some places in Nevada are slightly paler than the average. This small reddish shrew is found all the way to the Pacific coast of California, Oregon, and Washington. In these coastal areas it is somewhat darker and sometimes a trifle larger than elsewhere. It intergrades with a somewhat larger, grayer shrew in the Sierra Nevada of California. Along the Wasatch front in Utah, this Great Basin shrew intergrades with the larger, grayer shrew of the Rockies. Owing to the abrupt change in elevation, the zone of intergradation is rather narrow horizontally. In the latitude of Salt Lake City, populations of intergrades occur at between 8,700 and 9,000 feet elevation. The lowland shrew occurs in the eastern part of the Snake River Plains, and along the valleys of the Bear and Salt rivers into Wyoming. Along the northern edge of the Snake River Plains and on the western edge of the mountains of central Idaho the transition from lowland to montane habitats is abrupt and in consequence the zone of contact between small and large shrews is narrow. In northern Idaho and northwestern Montana the transition from lowland to highland is more gradual. Tributaries of the Columbia River system, especially the Clark Fork, provide a path for movement of lowland forms into intermontane basins of western Montana. In addition, the vegetational zones are found at lower elevations, and there are boreal forests in the lowlands rather than only in the mountains as is the case in Utah and Colorado. In this area, therefore, the zone of intergradation between the smaller lowland shrew and the larger montane shrew is more gradual and gradually intergrading populations are found over a relatively large area. This has been well demonstrated for northwestern Montana by Clothier (1950). In southern British Columbia and northern Washington this shrew in the mountains is large and in the intermontane valleys is small. There is extensive interdigitation of valleys and mountain ranges, and, consequently, of life-zones in this region. In a few places, recognizably distinct populations of the vagrant shrew occur within a few miles of one another, but in other places there are populations of intergrades. West of the Cascades no evidence of intergradation has been found and the two kinds occur almost side by side and maintain their distinctness.

These Great Basin shrews dwell in hydrosere communities as do their Rocky Mountain counterparts. In this arid region such a habitat obviously is the only one habitable for a shrew of the *vagrans* group. These shrews often maintain their predilection for such habitats when they reach the Pacific coast, and are commonly found in such places as coastal marshes, marshy meadows, and streamsides, while the woodlands are inhabited by other species.

These small shrews of the Great Basin and the small vagrant shrews of the Pacific Coast were called *Sorex vagrans* by Jackson.

Summary of Geographic Variation

Large reddish shrews of the coast of California and southwestern Oregon become smaller and darker to the north. From southwestern British Columbia they again become larger as one proceeds northward along the coast to Wrangell, Alaska, and north of that they again become smaller. Moving inland from the coast the shrews become markedly smaller in Alaska and British Columbia. The smaller inland and montane form occurs south through the Rocky Mountains, becoming slightly smaller in central Montana, slightly larger in southeastern New Mexico, and slightly smaller in western New Mexico and in Arizona. This montane form intergrades with a smaller more reddish Great Basin shrew, the zone of intergradation roughly following the western slope of the Rocky Mountains. The Great Basin shrew occurs westward to the Pacific Coast; there the Great Basin shrew occurs with, although in part it is ecologically separated from, the large reddish coastal shrews.

There seems to be an intergrading chain of subspecies of one species, the end members of which (the small Great Basin form and the large coastal form) are so different in size and ecological niche that they are able to coexist without interbreeding. In southern British Columbia the morphological differences are not so marked as farther south along the Pacific Coast. There, in British Columbia, reproductive isolation is not complete and occasional populations of intergrades occur. In Montana extensive intergradation occurs in a broad zone of transitional habitat. Along the western edge of the Rockies from Idaho south to Utah the zone of transition from montane to basin habitat is sharp and the zone of intergradation, although present, is fairly narrow, perhaps because there is little intermediate habitat which logically might be expected to be most suitable for intergrading populations.

The oldest name applied to a shrew of the group under consideration is *Sorex vagrans* Baird, 1858, the type locality of which is Willapa Bay, Pacific County, Washington. The name applies to the small vagrant shrew of this area, rather than to the larger forest dweller which has been known as *Sorex obscurus*. The name *S. vagrans*, in the specific sense, must therefore apply to all the shrews discussed which have heretofore been known by the names *S. pacificus*, *S. yaquinae*, *S. obscurus*, and *S. vagrans*.

A situation such as the one here described where well differentiated end members of a chain of subspecies overlap over an extensive geographic range throughout the year without interbreeding--thus reacting toward one another as do full species--so far as I know has not previously been found to exist in mammals. The overlapping end-members of the chain of subspecies of *Sorex vagrans* really do coexist; specimens of the overlapping subspecies have been taken together at the same localities from California to British Columbia. I have taken a specimen of *S. v. vagrans* and several of *S. v. setosus* in the same woodlot at Fort Lewis, Pierce County, Washington. Two subspecies of deer, *Odocoileus hemionus*, in the Sierra Nevada of California, occur together over a sizeable area but for only a part of each year that does not include the breeding season (Cowan, 1936:156-157). In the deer mouse, *Peromyscus maniculatus*, the geographic ranges of several pairs of subspecies meet at certain places without intergradation of the two kinds. In these instances well marked ecological differences exist between the subspecies involved. In western Washington, for example, the geographic range of the lowland subspecies, *P. m. austerus*, interdigitates to the east and west with the range of the montane and coniferous forest-inhabiting subspecies, *P. m. oreas*, and the two kinds have not been shown to intergrade. *Peromyscus maniculatus artemesiae* and *P. m. osgoodi* come together without interbreeding in Glacier National Park, Montana. *P. m. artemesiae* is almost entirely a forest-dwelling subspecies, whereas *osgoodi* is an inhabitant of open country. The two kinds do not actually occur together ecologically although they occur together in buildings at the edge of the woods (A. Murie, 1933:4-5).

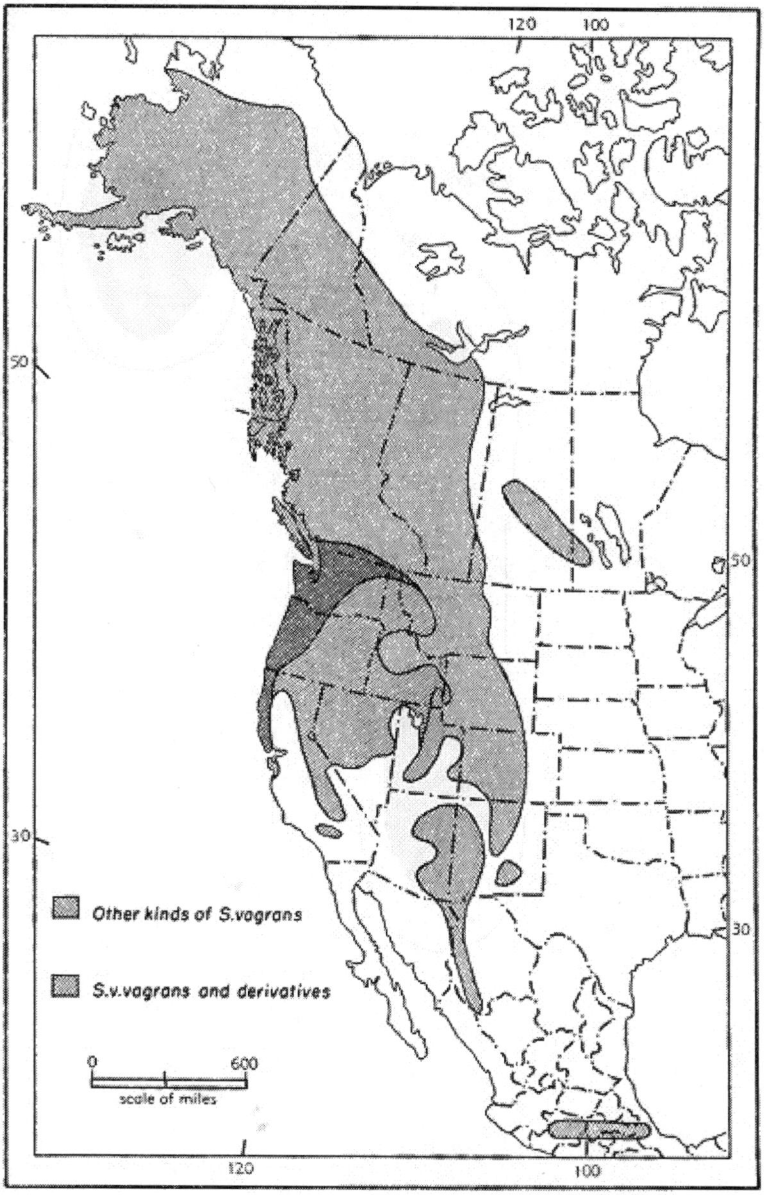

FIG. 5. Probable present geographic distribution of *Sorex vagrans*. The range of *S. v. vagrans* and its derivatives *S. v. vancouverensis*, *S. v. halicoetes*, and *S. v. paludivagus*, is shown by lines slanting in a different direction than those which mark the range of all the other subspecies of *S. vagrans*. The region in which *S. v. vagrans* occurs together with other

subspecies of *S. vagrans* is shown by the superposition of one pattern upon the other.

Cases of sympatric existence of two subspecies of one species are known in birds and in reptiles. Notable examples are in the gull, *Larus argentatus* (Mayr, 1940), in the Old World warbler, *Phylloscopus trochiloides* (Ticehurst, 1938), and in the great titmouse, *Parus major* (Rensch, 1933), of the Old World. In the first species the two end-members, the herring gull and the lesser black-backed gull, occur together over an extensive region from northern Europe and the British Isles throughout Fennoscandia. Fitch (1940) described a rassenkreis with overlapping subspecies in the garter snake *Thamnophis ordinoides*.

The geographic distribution of the species *Sorex vagrans* is shown in figure 5. The geographic range of the Great Basin subspecies is shown by a different pattern of lines than the other subspecies of *S. vagrans*. In the region in which the geographic range of the Great Basin subspecies overlaps those of the subspecies of the Pacific Coast, the pattern of shading for the Great Basin subspecies is superimposed on the patterns for the other subspecies.

ORIGIN OF THE *SOREX VAGRANS* RASSENKREIS

The distribution of the species *Sorex vagrans* and that of its immediate ancestors obviously has not always been the same; during glacial ages much of the present range of the species in Canada and in some of the higher mountains of the United States was covered with ice and not available to the shrew. Furthermore, large areas that are now too hot and dry to permit the existence of *S. vagrans* were at one time habitable. If we are to speculate on the manner in which the *Sorex vagrans* rassenkreis originated we must inquire into the nature and extent of these climatic changes.

The most recent epoch of geological time, the Pleistocene, is known to have been divided into a series of alternating glacial and interglacial ages. During the glacial ages continental and montane glaciers are judged to have covered much of Canada and the northern United States. Concurrently the major storm tracks of the west probably were shifted southward; in any event much of the now arid intermontane west was much better watered than it is today. The increased precipitation, and probably glacial meltwater, formed large lakes in the closed basins of the Great Basin. There were boreal forests at lower elevations than there are today in comparable latitudes and continuous boreal habitat probably connected many of the isolated mountain ranges of the southwest. That probability is supported by the presence of boreal animals and plants on many of these isolated ranges today. A boreal tree squirrel, such as *Tamiasciurus,* could hardly be suspected of crossing a treeless, intermontane desert valley, miles wide.

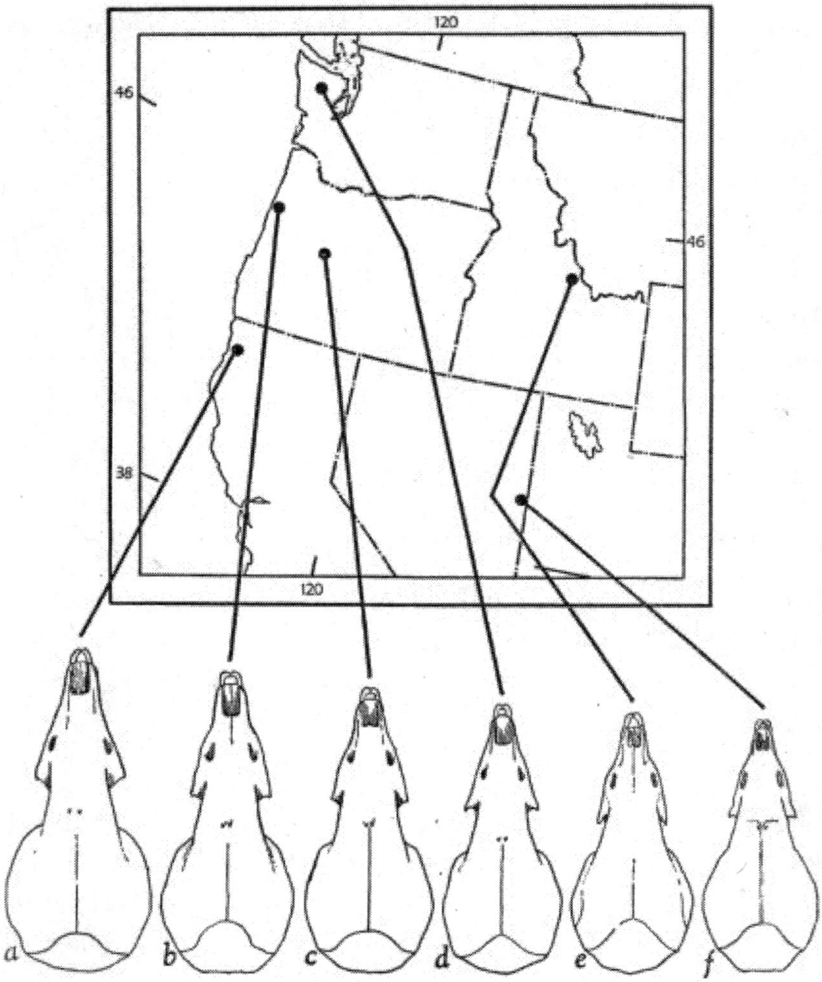

FIGS. 6*a*-6*f*. Fig. 6*a*. *Sorex vagrans* pacificus, 1 mi. N Trinidad, Humboldt Co., California, FC 1442. Fig. 6*b*. *S. v. yaquinae*, Newport, Lincoln Co., Oregon, AW 707. Fig. 6*c*. *S. v. yaquinae* (near *bairdi*), McKenzie Bridge, Lane Co., Oregon, AW 82. Fig. 6*d*. *S. v. setosus*, Reflection Lake, Jefferson Co., Washington, CMNH 4275. Fig. 6*e*. *S. v. obscurus*, 10 mi. SSW Leadore, Lemhi Co., Idaho, FC 1499. Fig. 6*f*. *S. v. vagrans*, Baker Creek, White Pine Co., Nevada, 88042 (after Hall, 1946:113).

Interglacial ages were characterized by warmth and aridity as compared to the glacial ages. Glaciers retreated or disappeared, boreal forests became montane in much of the United States, and the lakes in the Great Basin were reduced or disappeared. One can envision that during such times boreal mammals were isolated, their geographic ranges were restricted, and Sonoran mammals expanded their ranges.

Evidence is more extensive concerning the number and extent of glacial ages in the eastern than in the western part of North America. This evidence suggests a division of the Pleistocene into four glacial ages and four interglacial ages, the fourth interglacial age corresponding to the present time. More information is available about the Wisconsinan, or last, glacial age, than about the earlier ones, because the last glaciation in many montane areas destroyed evidence of earlier glaciations. The names of currently recognized glacial and interglacial ages of the Pleistocene are listed below. The names of interglacial ages are in Italic type.

Wisconsinan
Sangamonian
Illinoian
Yarmouthian
Kansan
Aftonian
Nebraskan

We may think of these ages as an alternating series of cool moist and warm dry periods during which boreal mammals, and other organisms, alternately moved southward (disappearing in the glaciated regions) and northward into previously glaciated areas (while disappearing from southern areas except on isolated mountain ranges). *Sorex vagrans* probably followed this pattern of movement and now is restricted to forested or well-watered places.

One possible series of events culminating in the formation of the *Sorex vagrans* rassenkreis may be thought of as having begun during the Illinoian age. With much of Canada, and perhaps also many areas in the Rockies, Cascades, and the Sierra Nevada covered with glacial ice, the shrew-stock ancestral to *Sorex vagrans* may well have occupied a more or less continuous range over the Colorado Plateau, the Columbian Plateau, the Great Basin, and in the forests of the Pacific Coast (as well as over part of eastern United States, as will be explained beyond; see fig. 7). At that time the species probably was a continuously interbreeding unit.

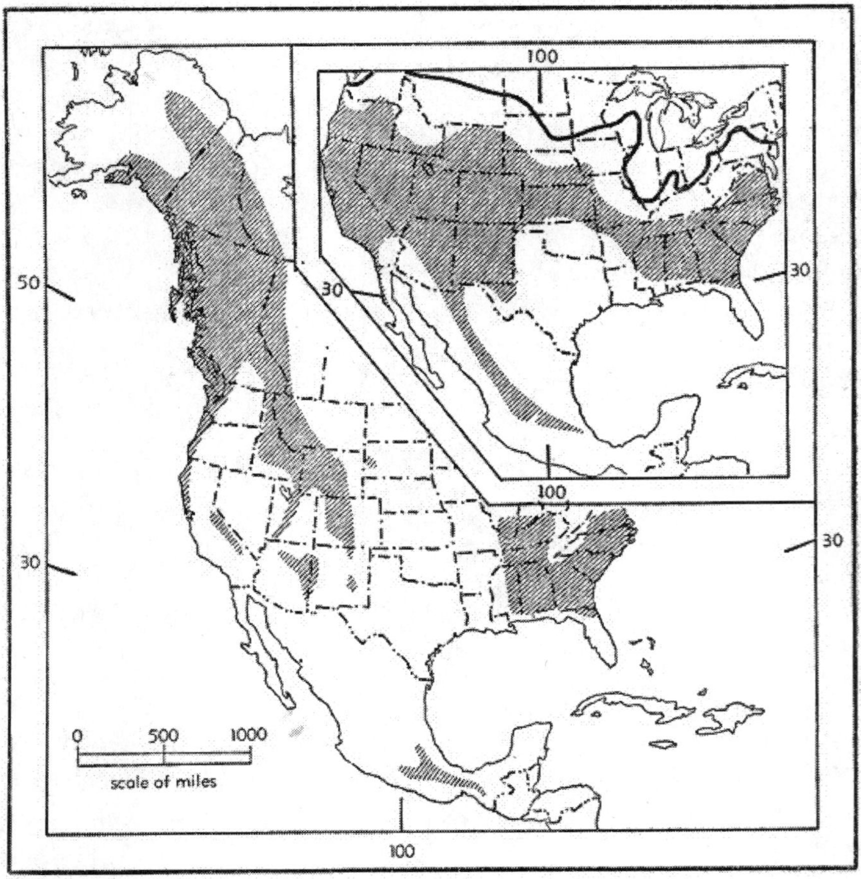

FIG. 7. Possible distribution in Illinoian (inset) and Sangamonian times of the ancestor of the *Sorex vagrans-ornatus-longirostris-veraepacis* complex. Approximate southern boundary of Illinoian glaciation marked by heavy line.

In the ensuing Sangamonian interglacial age all glaciers retreated or disappeared thereby opening up extensive areas in the north and in the higher mountains which were occupied by a boreal fauna, including *S. vagrans*. Concurrently the Great Basin, and probably also much of the Columbian Plateau, became dry, and desert conditions developed, perhaps much as they are today. Increasing aridity eliminated shrew habitat in most places between the Rocky Mountains and the Sierra Nevada-Cascade mountain chain with the result that the geographic range of the species resembled an inverted "U", one arm lying along the

Rocky Mountains and the other along the Cascade-Sierra Nevada axis; the connection between the two arms was in British Columbia (see fig. 7). At present *Sorex vagrans* does occur in isolated places in the Great Basin, but its existence there is tenuous and seemingly dependent upon the occurrence of permanent water such as Ruby Lake and Reese River. With such an arrangement as this it can readily be seen that gene flow between the eastern and western arms of the "U" would be greatly reduced by distance; consequently differentiation between the two might be expected.

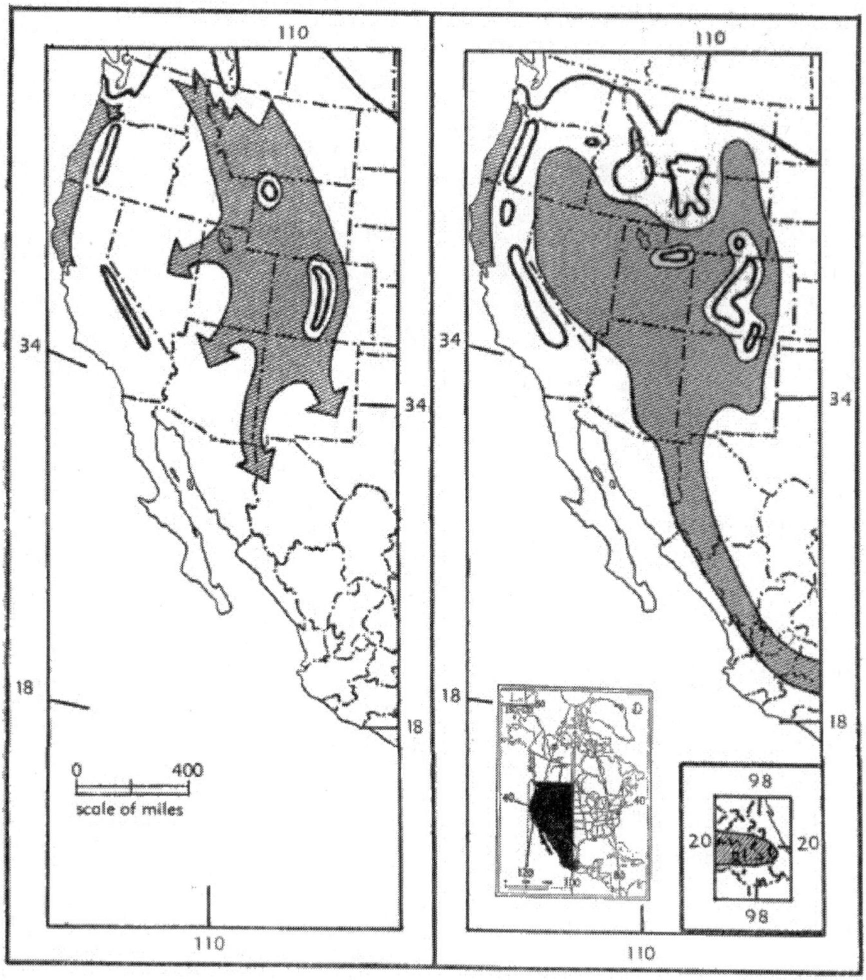

FIG. 8. Possible distribution of *Sorex vagrans* at two different times in the Wisconsinan Age. Left, early Wisconsinan; right, mid-Wisconsinan.

Wisconsinan glaciation again rendered Canada uninhabitable, and it is quite possible that extensive areas in the Rocky Mountains, the Cascades and the Sierra Nevada were heavily glaciated. With the elimination of the northern part of the "U", the eastern and western arms became isolated, if not by the width of the Columbian Plateau at least by the glaciated Cascade Mountains. At the same time extensive areas on the Colorado Plateau and much of the area south to the Mexican highlands were again occupied by the species. Finally the Great Basin, again being well-watered, provided suitable habitat for, and was reoccupied by, *Sorex vagrans* (see fig. 8). This reoccupation of the Great Basin took place probably from the Colorado Plateau and mountains of Arizona and Utah, since the present day shrews of the species *S. vagrans* in the Great Basin closely resemble Rocky Mountain shrews but differ markedly from the large endemic subspecies of the Pacific Coast.

Finally, with the waning of Wisconsinan ice, the species again was able to occupy northern and montane areas as it had during Sangamonian times. Again dessication of the Great Basin caused drastic restriction of shrew habitat. The small, marsh-dwelling kind of wandering shrew which had developed there around the lakes of Wisconsinan time occupied suitable habitat all the way to the Pacific coast where its range came into contact with that of the western arm of the Sangamonian "U."-pattern of shrew distribution (see fig. 9). The animals of this western segment and the new arrivals from the east were by this time so different from one another that the two kinds lived in the same areas without interbreeding. The descendants of the original western arm now are known as *Sorex vagrans sonomae, S. v. pacificus, S. v. yaquinae,* and *S. v. bairdi*. The newcomers from the east are known as *S. v. vagrans, S. v. halicoetes, S. v. paludivagus* and *S. v. vancouverensis*.

In addition to occupying the Pacific Coast from San Francisco Bay north to the Fraser Delta, the Great Basin subspecies populated the Columbia Plateau and the western foothills of the central and northern Rockies. By so doing that subspecies came into secondary contact with its own parent stock with which it was still in reproductive continuity in Utah. In some places in British Columbia differentiation between the two kinds had proceeded to such an extent that some reproductive isolation was effected, but in many other places the two interbred. The Rocky Mountain form spread north and west and occupied the Cascades and coastal lowlands in southwestern British Columbia and in Washington. Here the differentiation between the Rocky Mountain subspecies and the Great Basin subspecies was great enough to cause complete reproductive isolation.

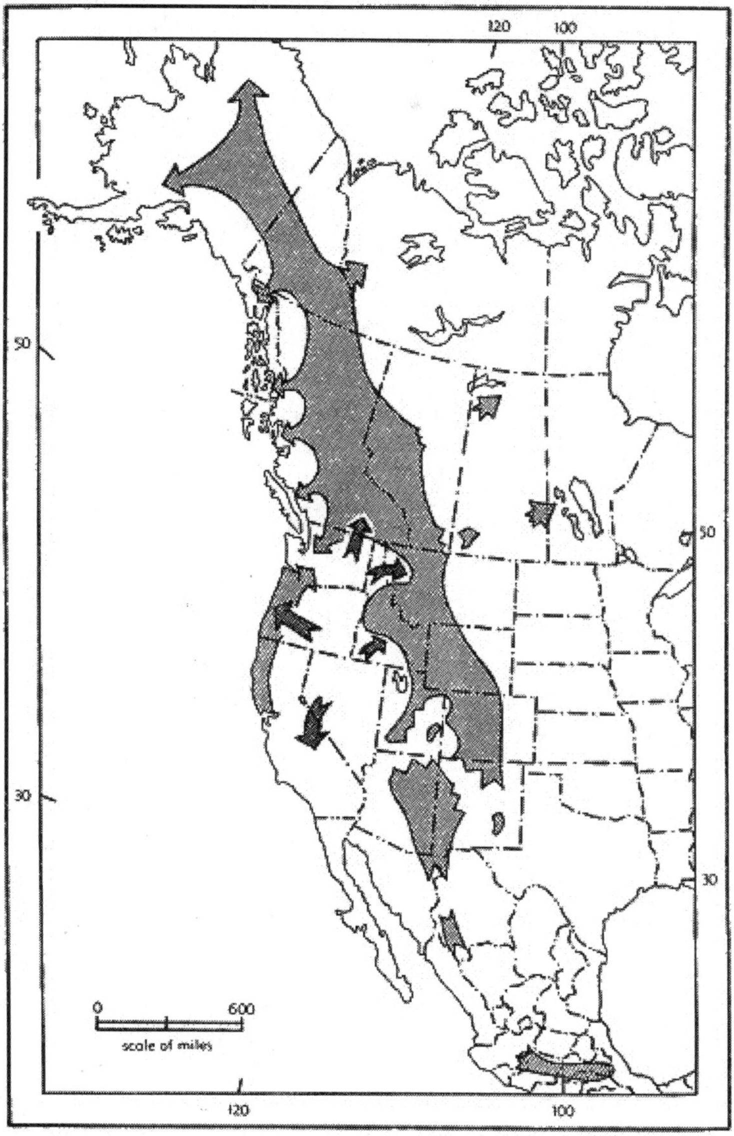

FIG. 9. Probable changes in the distribution of *Sorex vagrans* concurrent

with and following the dissipation of Wisconsinan ice. Dark arrows in Washington, Idaho, Oregon, and California, shows *S. v. vagrans*.

Deglaciation of the Sierra Nevada opened it up for reoccupation from the east by *Sorex vagrans*s of the Great Basin. In response to the montane environment the subspecies *obscuroides,* resembling the subspecies *obscurus* of the Rockies, developed.

Desiccation of the intermontane parts of New Mexico, Arizona, and Chihuahua, left "marooned" populations of *Sorex vagrans* on suitable mountain ranges. In this way *Sorex vagrans orizabae* may have been isolated in southern Mexico. The isolated populations of Arizona and New Mexico differentiated *in situ* into the subspecies *monticola* and *neomexicanus.*

Western Canada and Alaska were populated by shrews which originated in the habitable parts of the Rocky Mountains and Colorado Plateau during Wisconsinan time (as opposed to shrews originating, as subspecies, in the Great Basin or on the Pacific Coast). These shrews differentiated into the currently recognized subspecies of the west coast and coastal islands of British Columbia and Alaska in response to the different environments in these places, many of which were isolated; the subspecies *isolatus, mixtus, setosus, longicauda, elassodon, prevostensis, malitiosus,* and *alaskensis* are thought to have originated in this fashion after the areas now occupied by them were freed of Wisconsinan ice.

This group of shrews from the Rocky Mountains probably came into contact with the Pacific coastal segment of the species somewhere in northwestern Oregon. The clinal decrease in size from *S. v. pacificus* to *S. v. setosus* seems steepest in this area. Upon the establishment of this contact reproductive continuity was resumed, probably because the temporal separation of the two stocks involved was not so great as, say, that between *S. v. vagrans* and *S. v. pacificus,* and in addition the morphological differentiation was not so great.

On the eastern side of the Rockies the montane stock moved northeastward, occupying suitable territory opened up by the dissolution of the Laurentide ice sheet. Still later changes in the character of the northern plains owing to desiccation divided the range of the species and isolated *S. v. soperi* in Manitoba and central Saskatchewan and a population of *S. v. obscurus,* in the Cypress Hills. A number of semi-isolated stocks in central Montana became differentiated as a recognizable subspecies there.

A number of other boreal mammals have geographic ranges which resemble that of *Sorex vagrans,* except that the geographic ranges of subspecies do not overlap. Because of the general similarities of these geographic ranges, it is pertinent to examine the reasons suggested by students to account for the present geographic distributions of some of these other boreal species.

The red squirrel genus, *Tamiasciurus,* has a Rocky Mountain (and northern coniferous forest) species, *T. hudsonicus,* that occurs all along the Rocky Mountain chain and northward into Alaska. In the Cascade Mountains of Washington and British Columbia this species meets the range of a well marked western species, *T. douglasii,* with no evidence of intergradation. Dalquest (1948:86) attributes the divergence of the two species to separation in a glacial age but feels that the degree of difference between the two is too great to have all taken place during the Wisconsinan. Perhaps he has overemphasized the importance of the differences between the two, but, be that as it may, it seems that the two kinds differentiated during a glacial age when they were isolated, perhaps by ice on the Cascades into a coastal population and an inland population. One difference between the distribution of the red squirrels and vagrant shrew is that the squirrel of the Sierra Nevada is the species of the Pacific Coast, whereas the vagrant shrew of the Sierra Nevada was derived from the Great Basin population, which in turn was derived from the Rocky Mountain kind. Red squirrels do not occur on any of the boreal montane "islands" of Nevada. During the pluvial periods when hydrosere-loving shrews populated the Great Basin, that region may have been a treeless grassland. Vagrant shrews, then as now, probably depended on hydrosere communities, while red squirrels required trees. Therefore the shrews were able to traverse the Great Basin, while the Sierran red squirrels were of necessity derived from the coastal population.

The ecological requirements of jumping mice, genus *Zapus,* and the subspecies of *Sorex vagrans* that dwell in hydroseres are essentially similar. The species *Zapus princeps* lives in the Rocky Mountains, the Great Basin, the Sierra Nevada, and north to Yukon (Krutzsch, 1954:395). Its geographic range is similar to that of the montane and basin segments of *S. vagrans*. The species *Z. trinotatus* occurs along the Pacific coast and in the Cascades north to southwestern British Columbia. Its distribution thus coincides in general with that of the large red coastal subspecies of *S. vagrans*. Krutzsch (1954:368-369) thought that these two kinds of jumping mice were first separated by the formation of the Cascade

Mountains and the Sierra Nevada and finally by Pleistocene glaciation. The Sierran jumping mouse (*Zapus princeps*), as is the Sierran vagrant shrew, is more closely related to the jumping mouse of the Great Basin and of the Rocky Mountains than it is to the jumping mouse (*Z. trinotatus*) of the Pacific Coast, just as the Sierran vagrant shrew is related to the shrew of the Great Basin and Rocky Mountains. The jumping mouse also is limited in its distribution by hydrosere communities, not by forests.

In western North America there are two species of water or marsh shrews: *Sorex palustris* and *S. bendiri*. They have been placed in separate subgenera, but, as pointed out beyond, are closely related and here are placed in the same subgenus. The species *palustris* is found throughout the Rocky Mountains, north into Alaska, across the Great Basin into the Sierra Nevada, and west to the Pacific coast in Washington. The species *bendiri* is found from northwestern California north along the Pacific coast to southwestern British Columbia and east to the Cascades. Where the ranges of the two species overlap in western Washington they do not interbreed so far as is known, and are somewhat different in their ecology, *bendiri* being a lowland, and *palustris* being a montane, species. The two species probably were separated in a glacial period as seems to have been the case with the wandering shrews. Also, the water shrew of the Sierra Nevada is derived from that of the Great Basin and Rocky Mountains. *Sorex palustris* is tied closely in its distribution to hydrosere communities and is not dependent upon the presence of forests.

Red-backed mice, genus *Clethrionomys*, occur throughout the Rocky Mountains and west to the Cascades in Washington as the species *C. gapperi*. The species *C. californicus* is found along the Pacific Coast from California north to the Olympic Peninsula. Where the ranges of the two species meet in Washington they seem not to intergrade. In some glacial interval these two species may have evolved in the same manner as has been described for the species of *Zapus* and those of *Tamiasciurus*. No *Clethrionomys* are found in the Sierra Nevada, nor are red-backed mice found in the boreal islands of the Great Basin. It is not known why *Clethrionomys californicus* does not occur in the Sierra Nevada. Some boreal birds have distributional patterns similar to those of the mammalian examples cited above. One kind of sapsucker, *Sphyrapicus varius nuchalis*, occurs in the Rocky Mountains north into British Columbia and west to the Cascades and Sierra Nevada. A related kind, *S. varius ruber*, occurs along the Pacific Coast from California north into British Columbia. Recently Howell (1952) has shown that some intergradation takes place between *ruber* and *nuchalis* in Washington and British Columbia, although they do not intergrade freely. Previously the two kinds were thought not to intergrade and were regarded as two species. The two kinds intergrade also in northeastern California, although in that state *S. v. daggeti*, rather than *S. v. ruber*, is involved in the intergradation. Howell considered the two kinds to be conspecific with one another as well as with the eastern *S. varius*. He attributed a measure of the distinctness of *nuchalis* and *ruber* to their separation during a glacial period, but felt that the separation was much older than Wisconsinan. Whatever the time of separation, the pattern seems clear: *nuchalis* and *ruber* (as well as *varius*) were separated into montane, coastal, and eastern segments respectively, probably by glaciation (it seems to me in the Pleistocene), and have since re-established contact with one another.

The grouse genus *Dendrogapus* is divided into a Great Basin species, *D. obscurus*, which extends northward into British Columbia, and a Rocky Mountain species, *D. fuliginosus*, that is found in the Sierra Nevada and northward along the coast and Cascades into British Columbia. Although the two kinds have at times been considered conspecific, they differ in voice, hooting mechanism, and characters of the downy young, and so far no actual intergradation between the two has been shown (Grinnell and Miller, 1944:113). These grouse thus seem to offer additional evidence for a Pleistocene, possibly Wisconsinan, separation of the boreal fauna into a Rocky Mountain and a Pacific coastal segment.

A notable sidelight on these data is the frequency with which species in the Sierra Nevada have their closest relatives in the Rocky Mountains, rather than in the geographically nearer Cascades or coastal areas. This similarity in fauna of the Sierra Nevada and the Rockies was noted long ago by Merriam (1899:86).

RELATIONSHIPS WITH OTHER SPECIES

During the Sangamonian interval, isolated segments of the once widespread ancestral *Sorex vagrans* quite possibly persisted in such places as the Sierra Nevada, coastal southern California, the mountains of Arizona, New Mexico, and southern Mexico, and in the Black Hills (see fig. 6). One might expect that by Wisconsinan time these populations would have become reproductively isolated from their parent stock. They would therefore have remained specifically distinct when Wisconsinan *Sorex vagrans*, reoccupied these outlying areas, and may still be found isolated in places peripheral to the range of the ancestral species.

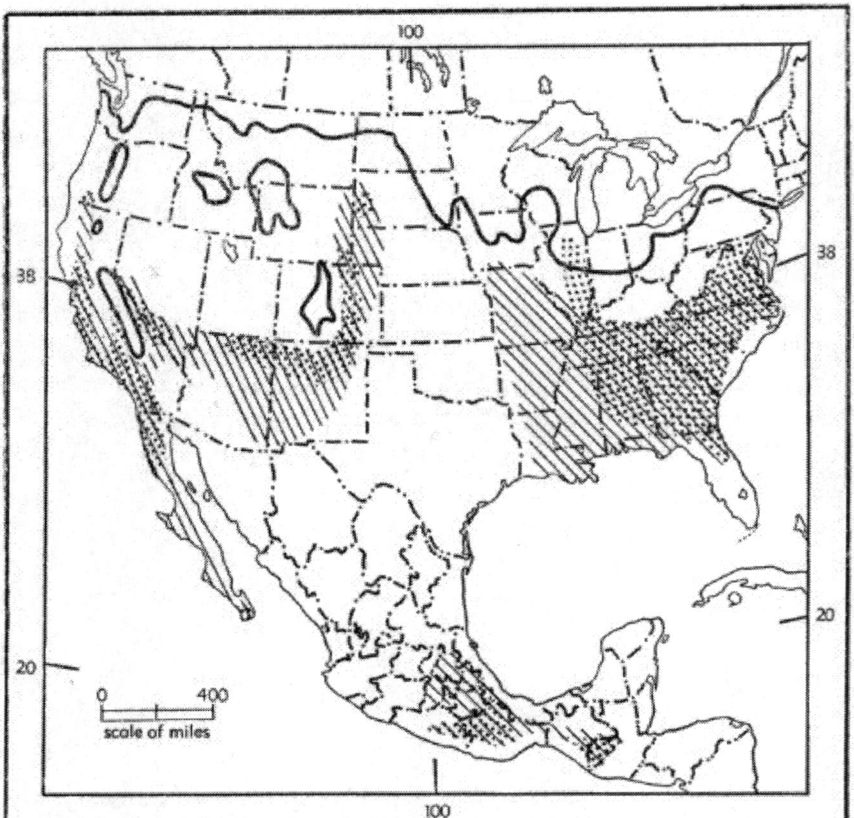

FIG. 10. Probable distribution of *S. veraepacis, S. longirostris,* and the *S. ornatus* group (stipple) and of their Wisconsinan ancestors (lines). Heavy line indicates limits of Wisconsinan glaciation.

In fact, we do find species closely related to *Sorex vagrans* in just such places today (fig. 10). Probably *Sorex ornatus,* including members of the *ornatus* group such as *S. trigonirostris, S. sinuosus, S. willeti, S. tenellus,* and *S. nanus,* and also *S. veraepacis,* arose by separation from the ancestral *vagrans* stock in Sangamonian time. Probably the eastern *S. longirostris* arose in a like manner. The ancestor of *S. ornatus* may have been isolated in southwestern California during Sangamonian time, spread north and south during the Wisconsinan age, and afterward given rise to *S. trigonirostris* and the modern *S. ornatus* complex of California and Baja California. In at least one place reproductive isolation between *ornatus* and the invading *S. vagrans* has broken down (Rudd, 1953); the place is a salt marsh along San Pablo Bay, where a hybrid population between *S. vagrans* and *S. sinuosus,* an *ornatus* derivative, has formed. *Sorex tenellus* may have been isolated in the Sierra Nevada in the Sangamonian interval, moved into the valleys east of the mountains during the Wisconsinan age, and become restricted to its present range since the retreat of the last ice. *Sorex nanus* may have occurred in the Black Hills and isolated mountains of Arizona and New Mexico during the Sangamonian interval and remained in these general areas during the Wisconsinan age. Its present range is peripheral to the main body of the Rockies and the Colorado Plateau.

The eastern species *Sorex longirostris* has many similarities with shrews of the *ornatus-vagrans* stock. *S. l. longirostris* is close in many ways to *S. nanus*. Indeed, the differences between the species *S. nanus, S. ornatus,* and *S. longirostris* seem to me to be of the same magnitude and indicate a similar period of differentiation from a common ancestor. The ancestor of *S. longirostris* may have gained access to the eastern United States in the Illinoian Age *via* the northern Great Plains south of the glacial boundary (fig. 7). The ancestor of *Sorex veraepacis* of southern Mexico probably reached that area in Illinoian time as part of the ancestral *vagrans* stock and probably attained its differentiation during the Sangamonian interval.

All the kinds of shrews so far discussed, including the *S. vagrans* complex, might thus be thought of as having had a common ancestor in the Illinoian Age. This entire group of shrews has the third unicuspid smaller than the fourth, a pigmented ridge from the apex to the cingulum of each upper unicuspid, and, in most individuals, lacks a post-mandibular foramen in the lower jaw (Findley, 1953:636-637). The pigment is not always prominent in *S. longirostris*.

Two other species of North American shrews, *Sorex palustris*, the water shrew, and *Sorex bendiri*, the marsh shrew, show these three characters to a greater or lesser degree, and it seems that these two species and the *vagrans-ornatus-veraepacis* group had a common ancestor, probably before Illinoian time for reasons stated beyond. I judge, however, that far from being subgenerically distinct as they have been considered to be, *S. palustris* and *S. bendiri* are actually closely related species of the same subgenus and may have differentiated from one another because of separation into eastern (*palustris*) and western (*bendiri*) segments in the Sangamonian interval, much as has been postulated concerning the eastern and western stocks of *Sorex vagrans*. Indeed, Jackson (1928:192) has noted that in the Pacific northwest the characters of the two kinds approach one another and become differences of degree only.

The widespread species *Sorex cinereus* resembles all the foregoing species in the ridges on the unicuspid teeth and in the lack of a post-mandibular foramen, but differs from those other species in having the third upper unicuspid larger than the fourth. The subspecies *S. cinereus ohionensis*, however, often has the sizes of these teeth reversed. With *S. cinereus* I include *S. preblei* (eastern Oregon) and *S. lyelli* (Sierra Nevada), both obviously closely related to *cinereus* as Jackson (1928:37) recognized when he included them in the *cinereus* group. *Sorex milleri* (Coahuila and central western Nuevo Leon) seems to me to resemble *S. cinereus* more than it does other species of North American *Sorex*, and I judge that it also belongs to the *cinereus* group. *Sorex cinereus* and its close relatives seem more closely related to the species which have thus far been discussed than they do to such other North American species as *S. arcticus*, *S. fumeus*, *S. trowbridgii*, *S. merriami*, and the members of the *S. saussurei* group; most of these five species last mentioned possess a post-mandibular foramen, lack pigmented unicuspid ridges, and have the third unicuspid larger than the fourth. Because of the morphological resemblances mentioned above, it seems likely to me that *S. cinereus* and the *vagrans-ornatus-veraepacis-palustris* complex had a common ancestor in early Pleistocene time. *Sorex cinereus* has recently been considered to be conspecific with the Old World *S. caecutiens* Laxmann (Van den Brink, 1953) which name, being the older, would apply to the circumpolar species.

Hibbard (1944:719) recovered *S. cinereus* and a species of *Neosorex* (a name formerly applied to the water shrew) from the Pleistocene (late Kansan) Cudahy Fauna. This indicates that the ancestors of the modern *S. cinereus* and of the water shrew had diverged from one another before that time. Brown (1908:172) recorded *S. cinereus* and *S. obscurus* from the Conard Fissure in Arkansas. These materials were deposited probably at a later time than was the Cudahy Fauna. The *S. obscurus* from Conard Fissure probably represents the ancestral *S. vagrans* stock which I think reached eastern United States in Illinoian time and gave rise to *S. longirostris*. The Conard Fissure material was deposited at a time (Illinoian?) when northern faunas extended farther south than they do today.

All of the species mentioned as having structural characters in common with *S. vagrans* seem to have arisen from a common ancestor which had already differentiated from the ancestor of such species as *S. arcticus*, *S. saussurei*, and others. Consequently all are here included in a single subgenus. The oldest generic name applied to a shrew of this group, other than the name *Sorex*, is *Otisorex* DeKay, 1842, type species *Otisorex platyrhinus* DeKay, a synonym of *Sorex cinereus*. The subgenus can be characterized as follows.

Subgenus **Otisorex** DeKay

1842. *Otisorex* DeKay, Zoology of New York, pt. 1, Mammalia, p. 22, and pl. 5, fig. 1. Type, *Otisorex platyrhinus* DeKay (= *Sorex cinereus* Kerr).

Third unicuspid usually smaller than fourth; upper unicuspids usually with pigmented ridge extending from apices medially to cingula, uninterrupted by antero-posterior groove; post-mandibular foramen usually absent. Includes the species *S. cinereus*, *S. longirostris*, *S. vagrans*, *S. ornatus*, *S. tenellus*, *S. trigonirostris*, *S. nanus*, *S. juncensis*, *S. willeti*, *S. sinuosus*, *S. veraepacis*, *S. palustris*, *S. bendiri*, *S. alaskanus*, and *S. pribilofensis*.

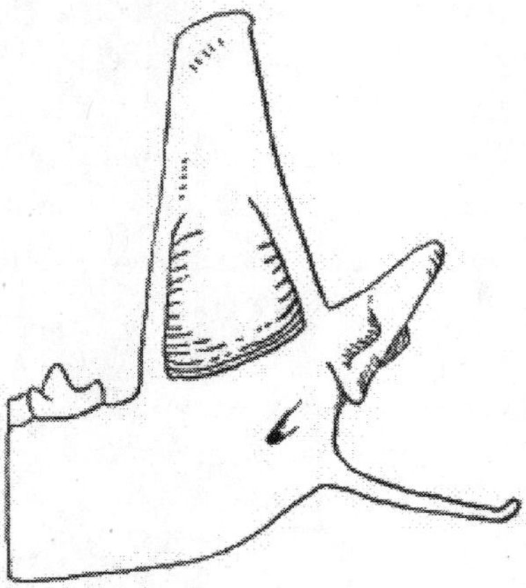

11

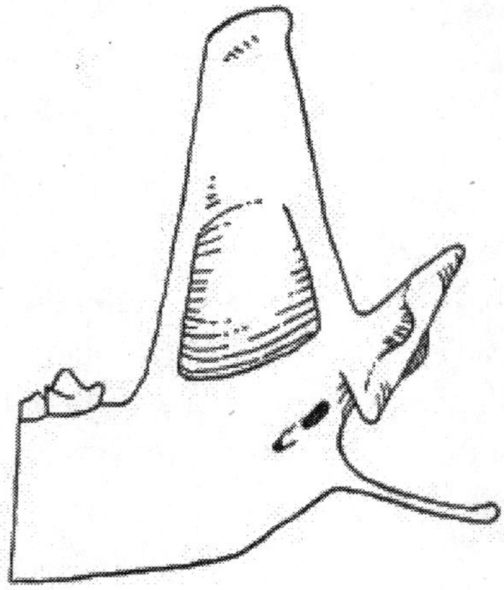

12

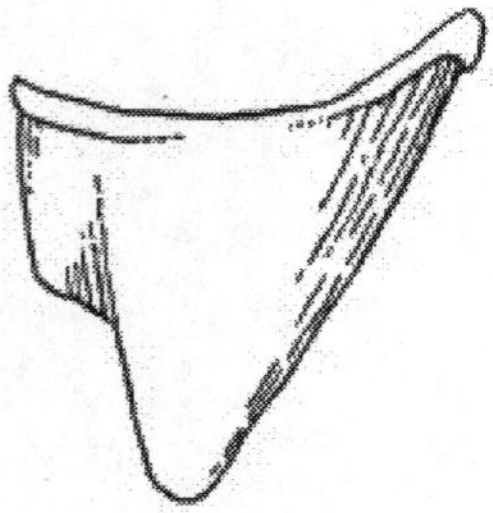

13

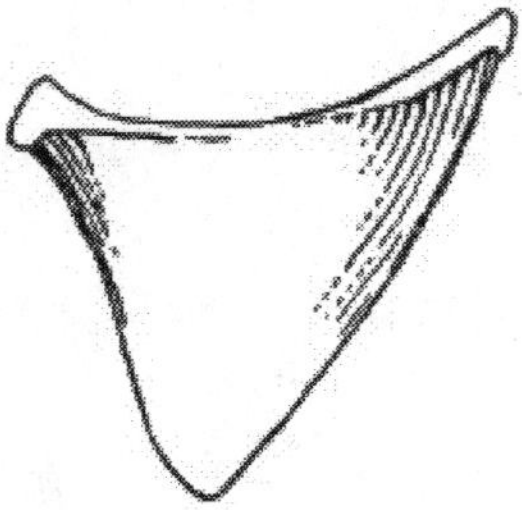

14

FIGS. 11-14.

Characters of the subgenera *Sorex* and *Otisorex*.

FIG. 11.

Medial view of right ramus of *Sorex (Otisorex) vagrans*. x 14.

FIG. 12.

Medial view of right ramus of *Sorex (Sorex) arcticus*. x 14.

FIG. 13.

Anterior view of left second upper unicuspid of *Sorex (Otisorex) vagrans*. x 45.

FIG. 14.

Anterior view of left second upper unicuspid of *Sorex (Sorex) arcticus*. x 45.

Other species of *Sorex* now occurring in North America differ from *Otisorex* in having the 3rd unicuspid usually larger than 4th, in lacking a pigmented ridge from the apices to the cingula of the upper unicuspids, and in usually possessing a well-developed post-mandibular foramen. Exceptions to the last mentioned character are *S. fumeus* and *S. dispar*. The subgenus *Sorex* in North America should include only the following species: *S. jacksoni, S. tundrensis, S. arcticus, S. gaspensis, S. dispar, S. fumeus, S. trowbridgii, S. merriami,* and all the members of the Mexican *S. saussurei* group.

The subgenera *Otisorex* and *Sorex* probably separated in early Pleistocene or late Pliocene. *Sorex* is unknown in North America earlier than the late Pliocene (Simpson, 1945:51).

In the genus *Microsorex* the characters of the subgenus *Otisorex* are carried to an extreme; the unicuspid ridges are prominent and end in distinct cusplets, and the 3rd unicuspid is not merely smaller than the 4th, but is reduced almost to the vanishing point. In addition, the post-mandibular foramen is absent. Although it is closer structurally to *Otisorex* than to *Sorex*, the recognition of *Microsorex* as a distinct genus seems warranted.

Figure 15 is intended to represent graphically some of the relationships discussed above. It must be re-emphasized that much of it is purely speculative, especially as regards actual time when various separations took place. It will be noted that I have indicated most separations as having taken place in interglacial ages. They are generally regarded as periods of warmth and aridity and, therefore, probably are times of segmentation of the ranges of boreal mammals and hence times exceptionally favorable to the process of speciation. Glacial ages, characterized by extensive and continuous areas of boreal habitat, probably were times of relatively unrestricted gene flow between many populations of boreal mammals and hence not favorable to rapid speciation.

Sorex vagrans

Wandering Shrew

The size of the wandering shrew varies from small in the subspecies *monticola* and *vagrans* to large in the subspecies *pacificus*. The tail makes up from a little more than a third to almost half of the total length. The color pattern ranges from tricolored through bicolored to almost monocolored. Color ranges from reddish (Sayal or Snuff Brown) to grayish in summer pelage and from black to light gray in winter. Diagnostic dental characters include: 3rd upper unicuspid smaller than 4th, and unicuspids, except 5th, with a pigmented ridge extending from near apex of each tooth medially to cingulum and sometimes ending as internal cusplet. *S. vagrans* differs from members of the *ornatus* group in less flattened skull, and in more ventrally situated foramen magnum that encroaches more on the basioccipital and less on the supraoccipital. The wandering shrew differs from *S. trowbridgii* and *S. saussurei* in the dental characters mentioned above. These dental characters also serve to distinguish *S. vagrans* readily from *S. cinereus*, *S. merriami*, and *S. arcticus* which may occur with *vagrans*. The large marsh shrew and water shrew, *S. palustris* and *S. bendiri*, can be distinguished at a glance from *S. vagrans* by larger size and darker color.

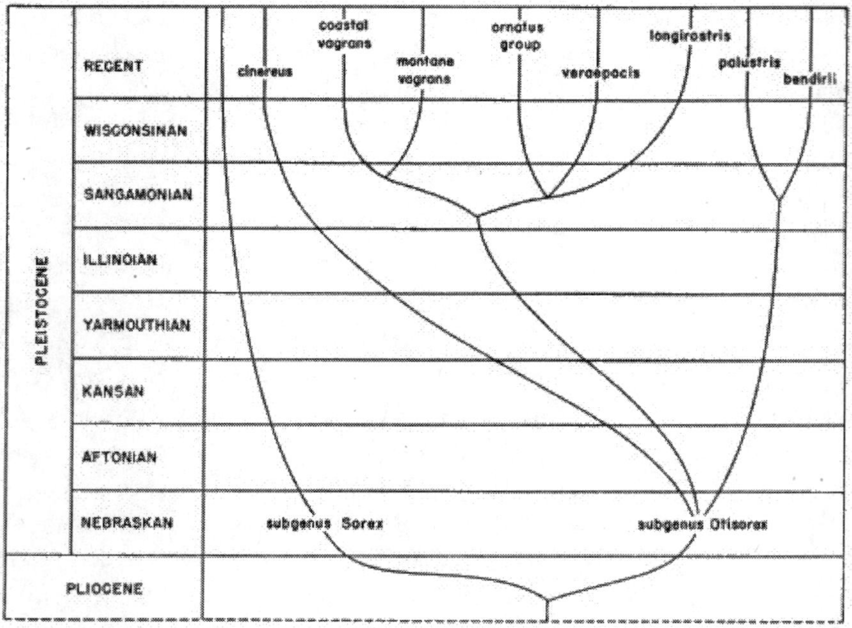

FIG. 15. Diagrammatic representation of the probable phylogeny of *Sorex vagrans* and its near relatives.

In the following treatment of the 29 subspecies of *Sorex vagrans*, the subspecies are arranged in geographic sequence, beginning with the southernmost large subspecies on the California coast and proceeding clockwise, north, east, south, and then west back to the starting point.

Sorex vagrans sonomae Jackson

Sorex pacificus sonomae Jackson, Jour. Mamm., 2:162, August 19, 1921.

Type.--Adult female, skin and skull; No. 19658, Mus. Vert. Zool.; obtained on July 2, 1913, by Alfred C. Shelton, from Gualala, on the Sonoma County side of the Gualala River, Sonoma Co., California.

Range.--Coastal California from Point Reyes north to Point Arena.

Diagnosis.--Size large; average and extreme measurements of 3 topotypes are: total length, 141.7 (141-143); tail, 59 (54-63); hind foot, 17 (17-17). Color reddish in summer, somewhat grayer in winter.

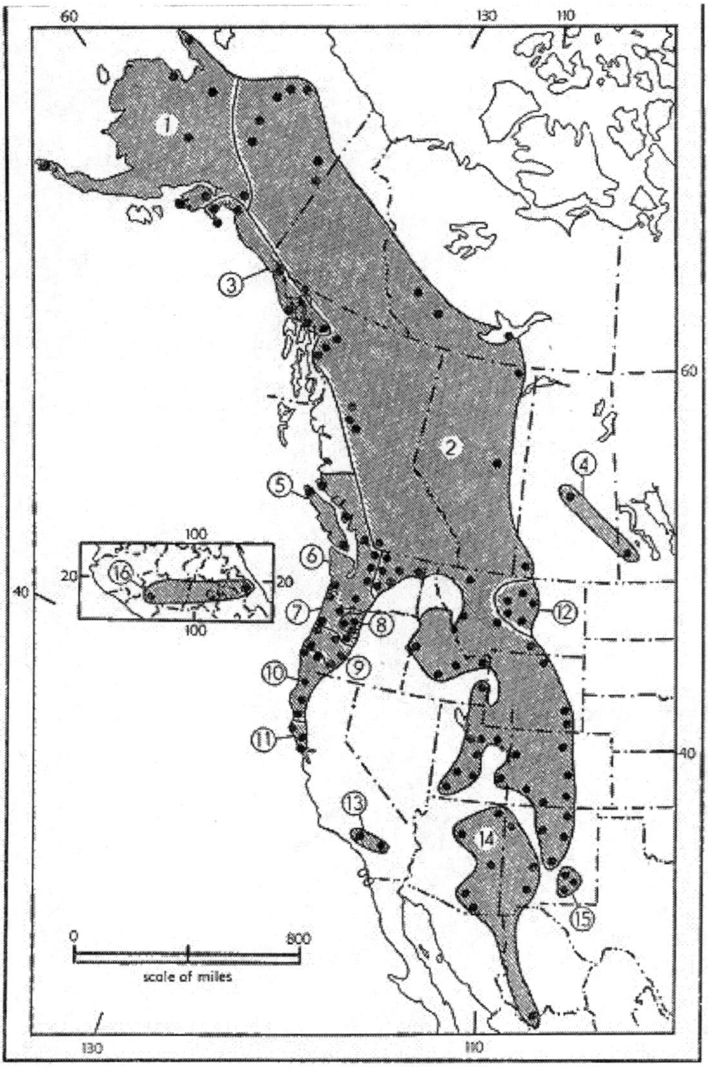

FIG. 16. Probable geographic ranges of 16 subspecies of *Sorex vagrans*.

Guide to subspecies

1. *S. v. shumaginensis*
2. *S. v. obscurus*
3. *S. v. alascensis*
4. *S. v. soperi*
5. *S. v. isolatus*
6. *S. v. setosus*
7. *S. v. bairdi*

Comparisons.--Differs from *S. v. pacificus*, with which it intergrades to the north, in average smaller size and somewhat darker color; differs from the sympatric *S. v. vagrans* in much larger size and more reddish color in both summer and winter.

Remarks.--This subspecies inhabits the Transition Life-zone below 300 feet, and occurs on moist ground in forests and beneath dense vegetation.

Marginal records.--CALIFORNIA: Point Arena (Grinnell, 1933:82); Monte Rio (Jackson, 1928:144); Inverness (Grinnell, 1933:82).

Sorex vagrans pacificus Coues

Sorex pacificus Coues, Bull. U. S. Geol. and Geog. Surv. Terr., 3 (3):650, May 15, 1877.

Sorex pacificus pacificus, Jackson, Jour. Mamm., 2:162, August 19, 1921.

Type.--Adult, sex unknown, skin and skull; No. 3266 U. S. Nat. Mus.; date of capture unknown; received from E. P. Vollum and catalogued on March 8, 1858; obtained at Ft. Umpqua, mouth of Umpqua River, Douglas Co., Oregon.

Range.--Coast of California and Oregon from Mendocino north to Gardiner.

Diagnosis.--Size large, largest of the species; average and extreme measurements of 8 specimens from Orick, Humboldt Co., California, are: total length, 143.1 (134-154); tail, 65.5 (59-72); hind foot, 17.5 (16-19). Color reddish in summer, browner or grayer in winter.

Comparisons.--See account of *S. v. sonomae* for comparison with that subspecies; averaging larger in all dimensions than *S. v. yaquinae* with which it intergrades to the north; much larger and has more reddish than the sympatric *S. v. vagrans*.

Remarks.--This subspecies occurs in the Canadian and Transition life-zones below 1500 ft. where there is found moist ground in or adjacent to heavy forests.

Specimens examined.--Total number, 76.

OREGON: *Douglas Co.*: Umpqua, 1 BS. *Coos Co.*: Marshfield, 1 BS; Myrtle Point, 1 BS. *Josephine Co.*: Bolan Lake, 1 SGJ.

CALIFORNIA: *Del Norte Co.*: Smith River, 2 BS; Gasquet, 4 BS; Crescent City, 17 BS. *Humboldt Co.*: Orick, 13 BS; 1 mi. N Trinidad, 18 FC; Trinidad

Head, 1 BS; Carson's Camp, Mad River, Humboldt Bay, 5 BS; Arcata, 3 BS; Cape Mendocino, 2 BS; 5 mi. S Dyerville, 1 BS. *Mendocino Co.*: Mendocino, 6 BS.

Marginal Records.--OREGON: Marshfield; Umpqua. CALIFORNIA: Gasquet; 5 mi. S Dyerville; Mendocino, thence up coast to point of beginning.

Sorex vagrans yaquinae Jackson

Sorex yaquinae Jackson, Proc. Biol. Soc. Washington, 31:127, November 29, 1918.

Sorex pacificus yaquinae, V. Bailey, N. Amer. Fauna, 55:364, August 29, 1936.

Type.--Adult female, skin and skull; No. 73051 U. S. Biol. Surv. Coll., obtained on July 18, 1895, by B. J. Bretherton, from Yaquina Bay, Lincoln Co., Oregon.

Diagnosis.--Size large for the species; average and extreme external measurements of 11 specimens from Oakridge, Lane Co., Oregon, are: total length, 125.3 (11-136); tail, 55.1 (49-61); hind foot, 14.9 (14-16). Color reddish in summer, browner or grayer in winter.

Comparisons.--See account of *S. v. pacificus* for comparison with that subspecies. Larger and more reddish than *S. v. bairdi* with which it intergrades to the north and east. Much larger and more reddish than the sympatric *S. v. vagrans.*

Remarks.--The name *yaquinae* actually applies to a population of intergrades between *pacificus* and *bairdi*. There is much variation over the range of the subspecies, and individuals from the western and southern parts are larger than those from the west slope of the Cascades. Specimens from Vida and McKenzie Bridge are smaller than those from Mapleton, Mercer, and the type locality but still seem closer to *yaquinae* than to topotypes of *bairdi*. Between Marshfield and Umpqua on the one hand, and the Columbia River and the Cascade Mountains on the other, the size of *Sorex vagrans* decreases quite rapidly from the large *pacificus* to the smaller *permiliensis*. Size decreases less rapidly northward along the coast than it does eastward toward the mountains; consequently, at any given latitude, coastal shrews are larger than mountain shrews. In this area of rapid change in size it is difficult to draw subspecific boundaries between *pacificus, yaquinae,* and *bairdi,* and this must be done somewhat arbitrarily.

Jackson (1928:141) remarked upon the possibility that intergradation between *pacificus* and *yaquinae* took place. He noted also the close resemblance between *yaquinae* and *bairdi*, and stated (*loc. cit.*) that specific affinity between the two might be demonstrated with more specimens. He had a series of eight specimens from Vida, Oregon, seven of which he assigned to *S. o. bairdi* and one to *yaquinae*. I have examined these specimens and find no more variation between the largest and the smallest than would be expected in any normally variable series of shrews. Vernon Bailey (1936:364) arranged *yaquinae* as a subspecies of *pacificus* without giving his reasons for so doing.

Specimens examined.--Total number, 65. OREGON: *Lincoln Co.*: type locality, 2 AW. *Benton Co.*: Philomath, 2 BS. *Lane Co.*: Mable, 1 OU; Vida, 4 BS, 1 OSC, 3 OU; McKenzie Bridge, 8 OSC, 3 AW, 17 OU, 2 SGJ; Mercer, 1 OSC, 1 OU; Mapleton, 3 BS; Oakridge, 11 OU. *Douglas Co.*: Gardiner, 2 BS; Elkhead, 1 BS. *Klamath Co.*: Crescent Lake, 3 OU.

Marginal Records.--OREGON: Yaquina Bay; *Philomath*; McKenzie Bridge; Prospect (Jackson, 1928:140); Crescent Lake; Gardiner.

Sorex vagrans bairdi Merriam

Sorex bairdi Merriam, N. Amer. Fauna, 10:77, December 31, 1895.

Sorex obscurus bairdi, Jackson, Proc. Biol. Soc. Washington, 31:127, November 29, 1918.

Type.--Adult female, skin and skull; No. 17414/24318, U. S. Biol. Surv. Coll.; obtained on August 2, 1889, by T. S. Palmer, from Astoria, Clatsop Co., Oregon.

Range.--Northwestern Oregon, south to Otis and east to Portland.

Diagnosis.--Size medium for the species; average and extreme external measurements of 6 specimens from the type locality are: total length, 126.3 (124-130); tail, 55.0 (52-57); hind foot, 15.0 (14-15). Color Fuscous to Sepia in summer, darker in winter, underparts buffy.

Comparisons.--For comparisons with *yaquinae* see account of that subspecies. More reddish and larger than *permiliensis* with which *bairdi* intergrades to the east; specimens from Portland show evidence of such intergradation. Some specimens from southern Tillamook County show an approach to *yaquinae*.

Remarks.--*S. v. bairdi* lives primarily in forests as do *yaquinae* and *pacificus*.

Specimens examined.--Total number, 39. OREGON: *Clatsop Co.*: type locality, 12 BS; Seaside, 3 BS. *Tillamook Co.*: Netarts, 1 OU; Tillamook, 2 OSC; Blaine, 1 AW; Hebo Lake, 1 SGJ; 5 mi. SW Cloverdale, 1 AW.

Multnomah Co.: Portland, 6 USNM. *Lincoln Co.*: Otis, 7 USNM; Delake, 1 KU. *Lane Co.*: north slope Three Sisters, 6000 ft., 4 BS.

Marginal Records.--OREGON: type locality; Portland; north slope Three Sisters; Taft (Macnab and Dirks, 1941:178).

Sorex vagrans permiliensis Jackson

Sorex obscurus permiliensis Jackson, Proc. Biol. Soc. Washington, 31:128, November 29, 1918.

Type.--Adult male, skin and skull; No. 91048, U. S. Biol. Surv. Coll.; obtained on October 2, 1897, by J. A. Loring from Permilia Lake, W base Mt. Jefferson, Cascade Range, Marion Co., Oregon.

Range.--The Cascade Mountains of Oregon from Mt. Jefferson north to the Columbia River.

Diagnosis.--Size medium for the species; average and extreme measurements of 14 specimens from the type locality are: total length, 117.7 (110-124); tail, 51.9 (45-58); hind foot, 14.0 (14-15). Pale reddish in summer, darker and brownish in winter.

Comparisons.--For comparison with *S. v. bairdi* see account of that subspecies. Larger than *S. v. setosus* except tail relatively shorter. More reddish in summer pelage than *setosus*.

Remarks.--*S. v. bairdi* is larger in the southern part of its range than elsewhere. Specimens from McKenzie Bridge, herein referred to *yaquinae*, are intermediate in character between *yaquinae* and *bairdi* or between *yaquinae* and *permiliensis*. The transition between *yaquinae* and *bairdi* is much more gradual than between *yaquinae* and *permiliensis*.

Specimens examined.--Total number, 21. OREGON: *Hood River Co.*: Mt. Hood, 2 BS. *Wasco Co.*: Camas Prairie, E base Cascade Mts., SE Mt. Hood, 1 BS. *Marion Co.*: Detroit, 1 BS; type locality, 17 BS.

Marginal Records.--OREGON: Mt. Hood; type locality; Detroit.

Sorex vagrans setosus Elliott

Sorex setosus Elliott, Field Columb. Mus. Publ. 32, zool. ser. 1:274, May 19, 1899.

Sorex obscurus setosus, Jackson, Proc. Biol. Soc. Washington, 31:127, November 29, 1918.

Type.--Adult male, skin and skull; No. 6213/238, Chicago Nat. Hist. Mus.; obtained on August 18, 1898, by D. G. Elliott from Happy Lake, Olympic Mts., Clallam Co., Washington.

Range.--Washington from the Cascades west; southwestern British Columbia west of 120deg W Longitude north to Lund.

Diagnosis.--Size medium for the species; average and extreme measurements of 20 specimens from the Olympic Mountains, Washington, are: total length, 117.3 (107-125); tail, 49.8 (41-54); hind foot, 13.4 (12-14). Color dark in both summer and winter.

Comparisons.--For comparison with *permiliensis* see account of that subspecies. Darker, longer-tailed, and somewhat larger cranially than *S. v. obscurus* with which it intergrades in southwestern British Columbia. Smaller in all dimensions, but much the same color as *S. v. longicauda* with which it intergrades along the British Columbian coast north of Lund. Larger, darker, less reddish, and longer-tailed than the sympatric *S. v. vagrans*.

Remarks.--*S. v. setosus* lives mostly in forests. According to Dalquest (1948:139) it is commonest at high altitudes in western Washington. In the Hudsonian Life-zone where shrew habitat is more restricted and marginal than it is at lower altitudes in the humid part of Washington, *setosus* might be expected to compete with *S. v. vagrans* and to supplant it. Records of occurrence in the Olympic Mountains suggest a degree of such separation there.

Specimens examined.--Total number, 135.

BRITISH COLUMBIA: Lund, Malaspina Inlet, 4 BS; Gibson's Landing, 10 BS; Port Moody, 19 BS; Langley, 2 BS; Chilliwack, 1 BS; Manning Park, 2 PMBC.

WASHINGTON: *Whatcom Co.*: Mt. Baker, 6 WSC; Barron, 1 BS. *Chelan Co.*: Clovay Pass, 1 WSC; Stehekin, 6 (4 WSC, 2 BS); Cascade Tunnel, 1 WSC. *King Co.*: Scenic, 1 WSC. *Kittitas Co.*: Lake Kachess, 1 WSC; Easton, 10 BS. *Clallam Co.*: 8 mi. W Sekin River, 1 WSC; mouth of Sekin River, 1 WSC; Clallam Bay, 1 CMNH; 7 mi. W Port Angeles, 1 WSC; Ozette Lake, 1 CMNH; 12 mi. S Port Angeles, 4 WSC; Forks, 1 CMNH; Deer Lake, 7 CMNH; Hoh Lake, 1 CMNH; Bogachiel Peak, 1 CMNH; Sol Duc Hot Springs, 3 CMNH; Sol Duc Park, 1 CMNH; Canyon Creek, 1 WSC; Sol Duc Divide, 2 WSC; Cat Creek, 2 WSC. *Jefferson Co.*: Jackson Ranger Station, 1 CMNH; Mt. Kimta, 2 CMNH; Reflection Lake, 6 CMNH; Blue Glacier, 3 CMNH. *Gray's Harbor Co.*: Westport, 1 WSC. *Pierce Co.*: Fort Lewis, 1 FC; Mt. Rainier, 19 (16 BS, 3 WSC). *Pacific Co.*: Tokeland, 2 BS. *Yakima Co.*: Gotchen Creek, 3 WSC; Mt. Adams, 1 WSC. *Skamania Co.*: Mt. St. Helens, 1.

OREGON: *Hood River Co.*: 2 mi. W Parkdale, 2 BS.

Marginal Records.--BRITISH COLUMBIA: Rivers Inlet (Anderson, 1947:20); *Agassiz* (Jackson, 1928:136); Chilliwack Lake. WASHINGTON: Barron; Lyman Lake (Jackson, 1928:137); Mt. Stuart (Dalquest, 1948:141); Mt.

Adams. OREGON: *2 mi. W Parkdale*. WASHINGTON: Ilwaco (Jackson, 1928:137); Lund, Malaspina Inlet.

Sorex vagrans longicauda Merriam

Sorex obscurus longicauda Merriam, N. Amer. Fauna, 10:74, December 31, 1895.

Type.--Adult male, skin and skull; No. 74711, U. S. Biol. Surv. Coll.; obtained on September 9, 1895 by C. P. Streator, from Wrangell, Alaska.

Range.--The British Columbian and Alaskan coasts from Rivers Inlet north to near Juneau and also certain islands including Etolin, Gravina, Revillagigedo, Sergeif, and Wrangell.

Diagnosis.--Size medium for the species, tail relatively long; average and extreme measurements of 17 specimens from the type locality are: total length, 128.4 (122-138); tail, 57.8 (53-66); hind foot, 15.1 (14-16). Color dark in summer and winter.

Comparisons.--For comparison with *S. v. setosus* see account of that subspecies. Larger and darker than *S. v. obscurus* with which it intergrades east of the humid coastal region; larger and darker than *S. v. alascensis* with which it intergrades in the Lynn Canal area; larger and darker than *S. v. calvertensis* which occurs on Calvert Island and Banks Island, British Columbia; differs from *S. v. insularis* of Smythe, Townsend, and Reginald islands in larger size and blackish rather than brown winter pelage; larger and relatively longer-tailed than *S. v. elassodon* which occurs on most of the islands west of the range of *longicauda*; larger and relatively longer-tailed than *S. v. isolatus*.

Specimens examined.--Total number, 151.

ALASKA: Wrangell, 54 BS; 8 AMNH; Crittenden Creek, 1 BS; Ketchikan, 2 BS; Loring, 11 BS.

BRITISH COLUMBIA: Port Simpson, 25 BS; Inverness, 15 BS; head of Rivers Inlet, 35 BS.

Marginal Records.--BRITISH COLUMBIA: Great Glacier, Stikine River (Jackson, 1928:133). ALASKA: Burroughs Bay (*ibid.*). BRITISH COLUMBIA: Bella Coola region (Anderson, 1947:19); head of Rivers Inlet; Spider Island (Cowan, 1941:101); Goose Island (Cowan, 1941:99); Princess Royal Island (Cowan, 1941:98); Pitt Island (*ibid.*); Metlakatla (Jackson, 1928:133); Port Simpson. ALASKA: Gravina Island (*ibid.*); Helm Bay (*ibid.*); Etolin Island (*ibid.*); Sergeif Island, mouth of Stikine River (*ibid.*); Sumdum Village (*ibid.*); Port Snettisham (*ibid.*).

Sorex vagrans mixtus Hall

Sorex obscurus mixtus Hall, American Nat., 72:462, September 10, 1938.

Type.--Adult male, skin and skull; No. 70376, Mus. Vert. Zool.; obtained on May 4, 1936, by R. A. Cumming, from Vanada, Texada Island, Georgia Strait, British Columbia.

Range.--Known only from the type locality.

Diagnosis.--Size medium; average and extreme measurements of 5 specimens from the type locality are: total length, 111 (108-117); tail, 48 (44-49); hind foot, 12 (12-13) (Hall, 1938:463).

Comparisons.--Color much as in *S. v. setosus* or *S. v. isolatus;* palate longer than that of *isolatus* or *setosus;* hind foot shorter than either; smaller than *S. v. longicauda.*

Sorex vagrans isolatus Jackson

Sorex obscurus isolatus Jackson, Jour. Washington Acad. Sci., 12:263, June 14, 1922.

Type.--Adult male, skin and skull; No. 177719, U. S. Biol. Surv. Coll.; obtained on May 21, 1911, by Alexander Wetmore from mouth of Millstone Creek, Nanaimo, Vancouver Island, British Columbia.

Range.--Vancouver Island.

Diagnosis.--Size medium; measurements of two from the type locality are: total length, 113, 118; tail, 48, 49; hind foot, 14, 14. Dark in summer and winter, underparts brownish.

Comparisons.--Smaller than *S. v. setosus* but color much the same; resembles *S. v. obscurus* in size and cranial characters but darker in all pelages; similar in color to *S. v. vancouverensis* with which *isolatus* is sympatric but with longer tail, longer hind feet, broader rostrum and larger teeth. For comparison with *S. v. mixtus* see account of that subspecies.

Remarks.--*S. v. isolatus* and *S. v. vancouverensis* seemingly approach one another morphologically more closely than do any other pair of sympatric subspecies of *Sorex vagrans.* The exceptions may be *S. v. vagrans* and *S. v. obscurus* which are geographically sympatric in a few places although they may be ecologically separated.

Specimens examined.--Total number, 9. BRITISH COLUMBIA, Vancouver Island: Nanaimo, 3 BS; Barclay Sound, 1 AMNH; Goldstream, 5 BS.

Marginal Records.--BRITISH COLUMBIA, Vancouver Island. (Anderson, 1947:19): Cape Scott; Victoria.

Sorex vagrans insularis Cowan

Sorex obscurus insularis Cowan, Proc. Biol. Soc. Washington, 54:103, July 31, 1941.

Type.--Adult female, skin and skull; No. 3110, Prov. Mus. British Columbia; obtained on August 24, 1938, by T. T. and E. B. McCabe from Smythe Island, Bardswell Group, British Columbia.

Range.--Smythe, Townsend, and Reginald islands, British Columbia.

Diagnosis.--Size medium; average and extreme measurements of 50 specimens from within the range of the subspecies are: total length, 122.3 (111-134); tail 52.6 (46-58); hind foot, 14.6 (13-15) (Cowan, 1941:107).

Comparisons.--Smaller externally and cranially than *S. v. longicauda* and brown instead of blackish or grayish in winter pelage. Skull broader than that of *S. v. calvertensis* and color brown rather than blackish or grayish in winter pelage.

Remarks.--*S. v. insularis* occurs together with *S. cinereus* on Townsend and Smythe islands. *S. vagrans* far outnumbered the cinereus shrew (Cowan, 1941:96).

Records of occurrence.--BRITISH COLUMBIA (Cowan, 1941:104): Smythe Island, Townsend Island, Reginald Island.

Sorex vagrans calvertensis Cowan

Sorex obscurus calvertensis Cowan, Proc. Biol. Soc. Washington, 54:103, July 31, 1941.

Type.--Adult male, skin and skull; No. 1947, Prov. Mus. British Columbia; obtained on July 14, 1937, by T. T. and E. T. McCabe from Safety Cove, Calvert Island, British Columbia.

Diagnosis.--Size medium; average and extreme measurements of 13 specimens from Calvert Island are: total length, 121.6 (109-129); tail, 54.0 (52-58); hind foot, 14.7 (13-15) (Cowan, 1941:106). Blackish or grayish in winter pelage.

Comparisons.--Smaller externally and cranially and paler in winter and summer than *S. v. longicauda*; for comparisons with *S. v. insularis* see account of that subspecies.

Remarks.--*S. v. calvertensis* seems to be the only shrew on Calvert and Banks islands.

Records of occurrence.--BRITISH COLUMBIA (Cowan, 1941:103): Safety Cove, Calvert Island; Larson Harbor, Banks Island.

Marginal Records.--BRITISH COLUMBIA: Larson Harbor, Banks Island; type locality.

Sorex vagrans malitiosus Jackson

Sorex obscurus malitiosus Jackson, Proc. Biol. Soc. Washington, 32:23, April 11, 1919.

Type.--Adult female, skin and skull; No. 8401; Mus. Vert. Zool.; obtained on May 21, 1909, by H. S. Swarth from east side of Warren Island, Alaska.

Range.--Warren and Coronation islands, Alaska.

Diagnosis.--Size medium; average and extreme measurements of 5 topotypes are: total length, 129.8 (126-135); tail, 56.4 (53-61); hind foot, 15.4 (15-16). Color brownish in summer, brownish rather than blackish in winter.

Comparisons.--Somewhat more brownish than *S. v. longicauda* but resembling it in size; skull slightly more flattened and rostrum broader. Larger than *S. v. elassodon*. Larger and relatively longer-tailed than *S. v. alascensis*.

Records of occurrence.--ALASKA (Jackson, 1928:130): Warren Island; Coronation Island.

Sorex vagrans elassodon Osgood

Sorex longicauda elassodon Osgood, N. Amer. Fauna, 21:35, September 26, 1901.

Sorex obscurus elassodon, Elliott, Field Columb. Mus. Publ. 105, zool. ser. 6:450, 1905.

Type.--Adult male, skin and skull; No. 100597, U. S. Biol. Surv. Coll.; obtained on June 13, 1900, by W. H. Osgood from Cumshewa Inlet, near old Indian village of Clew, Moresby Island, Queen Charlotte Islands, British Columbia.

Range.--Alaskan and British Columbian islands from Admiralty Island south to Moresby Island.

Diagnosis.--Size medium; average and extreme measurements of 4 from the type locality are: total length, 126, (119-131); tail, 53.5 (52-55); hind foot, 13.8 (13-14). Color dark.

Comparisons.--Smaller with relatively smaller tail and hind feet than *S. v. longicauda*, but resembling it in color. Smaller and paler than *S. v. prevostensis* with relatively narrower rostrum. Larger, darker, and with relatively longer tail than *S. v. obscurus*. Resembles *S. v. alascensis* but hind foot smaller and skull relatively narrower. Smaller than *S. v. malitiosus*.

Remarks.--In the northern part of its range *S. v. elassodon* occurs with *Sorex cinereus*. In the southern part it is the only shrew present.

Specimens examined.--Total number 93.

ALASKA: near Killisnoo, Admiralty Island, 2 BS; Kupreanof Island, 15 BS; Petersburg, Mitkof Island, 10 BS; Woewodski Island, 4 AMNH; Kasaan Bay, Prince of Wales Island, 18 BS.

BRITISH COLUMBIA: Cumshewa Inlet, Moresby Island, 25 BS; Massett, Graham Island, 6 BS; Queen Charlotte Islands, 13 AMNH.

Marginal Records.--ALASKA: Hawk Inlet, Admiralty Island (Jackson, 1928:131); Kupreanof Island; Mitkof Island; St. John Harbor, Zarembo Island (Jackson, 1928:131); Kasaan Bay, Prince of Wales Island; Duke Island (Jackson,

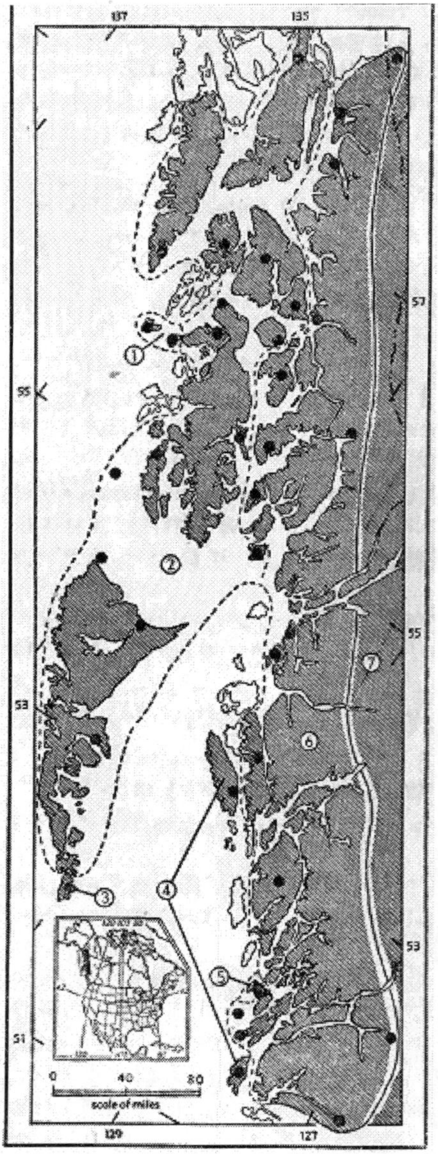

FIG. 17. Probable geographic ranges of the subspecies of *Sorex vagrans* on the coast of British Columbia and southeastern Alaska.

1. *Sorex vagrans malitiosus*
2. *Sorex vagrans elassodon*
3. *Sorex vagrans prevostensis*
4. *Sorex vagrans calvertensis*
5. *Sorex vagrans insularis*

1928:131). BRITISH COLUMBIA: Massett, Graham Island, Queen Charlotte Islands; type locality; Langara Island, Queen Charlotte Islands (Jackson, 1928:131). ALASKA: Forrester Island (*ibid.*); Rocky Bay, Dall Island (*ibid.*); Shakan (really on Kosciusko Island) (*ibid.*); Point Baker (*ibid.*); Kuiu Island (*ibid.*); Port Conclusion, Baranof Island (*ibid.*).

Sorex vagrans prevostensis Osgood

Sorex longicauda prevostensis Osgood, N. Amer. Fauna, 21:35, September 26, 1901.

Sorex obscurus prevostensis, Elliott, Field Columb. Mus. Publ. 105, zool. ser. 6:450, 1905.

Type.--Adult male, skin and skull; No. 100618, U. S. Biol. Surv. Coll.; obtained on July 3, 1900, by W. H. Osgood from north end of Prevost Island (Kunghit Island on some maps) on coast of Houston Stewart Channel, Queen Charlotte Islands, British Columbia.

Range.--Known only from the type locality.

Diagnosis.--Size medium; measurements of two specimens from the type locality are: total length, 132, 142; tail, 53, 59; hind foot, 14, 15. Color dark.

Comparisons.--Larger and darker than *S. v. elassodon*. Resembles *S. v. longicauda* but darker, tail relatively somewhat shorter on the average and rostrum relatively slightly broader.

Specimens examined.--Total number, 14. BRITISH COLUMBIA: Prevost Island, Queen Charlotte Group, 14 BS.

Sorex vagrans alascensis Merriam

Sorex obscurus alascensis Merriam, N. Amer. Fauna, 10:76, December 31, 1895.

Sorex glacialis Merriam, Proc. Washington Acad. Sci., 2:16, March 14, 1900, type from Point Gustavus, east side of entrance to Glacier Bay, Alaska.

S[orex]. alascensis, Merriam, Proc. Washington Acad. Sci., 2:18, March 14, 1900.

[Sorex glacialis] alascensis, Elliott, Field Columb. Mus. Publ. 45, zool. ser. 2:372, 1901.

Sorex alascensis alascensis, Miller, U. S. Nat. Mus. Bull., 79:16, December 31, 1912.

Type.--Adult female, skin and skull; No. 73539, U. S. Biol. Surv. Coll.; obtained on July 10, 1895, by C. P. Streator from Yakutat, Alaska.

Range.--The coast of southern Alaska from the vicinity of Juneau west to include eastern part of the Kenai Peninsula.

Diagnosis.--Size medium for the species; average and extreme measurements of 9 specimens from 9 mi. W and 4 mi. N of Haines, Alaska, are: total length, 110 (104-128); tail, 45.4 (41-52); hind foot, 14 (14-14). Color grayish brown.

Comparisons.--For comparison with *S. v. longicauda* and *S. v. elassodon* see accounts of those subspecies. Resembles *S. v. obscurus* in color but differs in larger skull, longer hind foot and in somewhat darker color. Larger and darker than *S. v. shumaginensis*; the two intergrade near the base of the Kenai Peninsula.

Remarks.--This subspecies is transitional between the large, usually dark subspecies of the southeastern Alaskan and British Columbian coast and islands, and the smaller, paler subspecies of western and interior Alaska. There seem to be no sharp breaks between *alascensis* and *shumaginesis*. North of Haines, Alaska, size of shrews decreases in a short distance across a narrow intergradational zone between *alascensis* and *obscurus*. Throughout most of its range *S. v. alascensis* occurs with *Sorex cinereus*.

Specimens examined.--Total number, 88.

ALASKA: Orca, 1 BS; Montague Island, Prince William Sound, 2 BS; Yakutat, 8 BS; north shore Yakutat Bay, 2 BS; Yakutat Bay, 1 BS; E side Chilkat River, 100 ft., 9 mi. W and 4 mi. N Haines, 12 KU; 1 mi. S Haines, 5 ft., 10 KU; 7 mi. SSE Haines, 10 ft., 2 KU; N end Sullivan Island, 10 ft., 6 KU; SE end Sullivan Island, 10 ft., 2 KU; Glacier Bay, 3 BS; Mendenhall River, 1 BS; Juneau, 36 BS.

BRITISH COLUMBIA: Sheslay River, 1 AMNH; headwaters Sheslay River, 1 AMNH.

Marginal Records.--ALASKA: Valdez Narrows, Prince William Sound (Jackson, 1928:128); north shore Yakutat Bay; east side Chilkat River, 100 ft., 9 mi. W and 4 mi. N Haines. BRITISH COLUMBIA: Sheslay River (Jackson, 1928:128). ALASKA: Juneau; Glacier Bay; Montague Island, Prince William Sound (ibid.); Port Nell Juan (ibid.).

Sorex vagrans shumaginensis Merriam

Sorex alascensis shumaginensis Merriam, Proc. Washington Acad. Sci., 2:18, March 14, 1900.

[*Sorex glacialis*] *shumaginensis,* Elliott, Field Columb. Mus. Publ. 45, zool. ser. 2:373, 1901.

Sorex obscurus shumaginensis, Allen, Bull. American Mus. Nat. Hist., 16:228, July 12, 1902.

Type.--Adult male, skin and skull; No. 97993, U. S. Biol. Surv. Coll.; obtained on July 17, 1899, by De Alton Saunders from Popof Island, Shumagin Islands, Alaska. (Measured by C. Hart Merriam and numbered 2210 in A. K. Fisher's catalog.)

Range.--Southwestern Alaska from Seward Peninsula southeasterly to western part of Kenai Peninsula and southwesterly to the southwestern end of the Alaskan Peninsula.

Diagnosis.--Size medium to small for the species; average and extreme measurements of 6 specimens from King Cove, Alaska, are: total length, 112.7 (107-118); tail, 48.3 (45-52); hind foot, 13.8 (13-14). Tending toward the development of a tricolor pattern, the back darkest, the sides buffy, and the venter paler.

Comparisons.--Paler and more definitely tricolored than *S. v. obscurus*; also with relatively shorter palate, narrower rostrum and smaller teeth. For comparison with *S. v. alascensis* see account of that subspecies.

Remarks.--*S. v. shumaginensis* occurs together with *Sorex cinereus* over much of southwestern Alaska. Part of the range of *shumaginensis* falls within the tundra of the Arctic Life-zone. This may be a partial explanation of the tricolored pattern of the animal. *Sorex tundrensis, S. cinereus ugyunak*, and *S. cinereus haydeni*, shrews which dwell mostly in treeless areas, are markedly tricolored, or bicolored. *Sorex arcticus*, however, although tricolored, is found in forested areas.

Specimens examined.--Total number, 340. ALASKA: Sawtooth Mts., Nome River, 2 AMNH; Nulato, 5 BS; St. Michaels, 1 BS; Bethel, 7 BS; Aniak, 1 BS; Skeventna River, 1 BS; 6 mi. WSW Snowshoe Lake, 1 KU; 1 mi. NE Anchorage, 1 KU; Tyonek River, 48 BS; Hope, 15 BS; Hope, Mts. near, 13 BS; Moose Camp, 3 AMNH; Kenai Peninsula, 24 AMNH; Kakwok River, 80 mi. up, 1 BS; Kakhtul River, 5 BS; Kakwok, 3 BS; Goodnews Bay, 1 BS; Lake Aleknagik, 6 BS; Nushagak River, 25 mi. above Nushagak, 1 BS; Dillingham, 1 BS; Nushagak Village, 15 BS; Homer, 1 AMNH; Kenai Mts., 37 AMNH; Seldovia, 24 AMNH; Barabor, 1 AMNH; Caribou Camp, 7 AMNH; Ugagik River, 3 BS; Becharof Lake, 8 BS; Cold Bay, 14 BS; Kanatak, Portgage Bay, 4 BS; Chignik, 6 BS; Moller Bay, 1 BS; Alaska Peninsula, near Popof Island, 6 AMNH; Frosty Peak, 15 BS; Morzhovoi Bay, 7 BS; Ungu Island, 3 BS; Sand Point, Popof Island, 45 AMNH; Popof Island, 3 BS.

Marginal Records.--ALASKA: Nome River; Nulato; Kuskokwim River, 200 mi. above Bethel, Crooked Creek (Jackson, 1928:126); 6 mi. WSW Snowshoe Lake; Seldovia; mts. near Hope; Morhzovoi Bay; thence along coast to St. Michael.

Sorex vagrans obscurus Merriam

Sorex vagrans similis Merriam, N. Amer. Fauna, 5:34, July 31, 1891, *nec. S. similis* Hensel, Zeitschr. der Deutsch. Geolog. Gesellsch., 7:459, 1855 (= *Neomys similis*).

Sorex obscurus Merriam, N. Amer. Fauna, 10:72, December 31, 1895, new name for *Sorex vagrans similis* Merriam.

Sorex obscurus obscurus, Miller, Bull. U. S. Nat. Mus., 79:15, December 31, 1912.

Type.--Adult female, skin and skull; No. 23525/30943, U. S. Biol. Surv. Coll.; obtained on August 26, 1890, by Vernon Bailey and B. H. Dutcher from near Timber Creek, 8200 ft., Lemhi Mts., 10 mi. SSW Junction (now Leadore), Lemhi Co., Idaho.

Range.--Mountainous interior of western North America from central Alaska east across Yukon and southwestern Northwest Territories to northeastern Alberta, south in the mountains through north-central and western Washington, Idaho, western Montana, Wyoming, Utah, and Colorado, into northern New Mexico.

Diagnosis.--Size medium to small for the species; average and extreme measurements of 9 topotypes are: total length, 110.3 (105-117); tail, 46.4 (42-50); hind foot, 13.1 (12.5-13.5). Color grayish or brownish gray in summer, light grayish in winter.

Comparisons.--For comparisons with *S. v. setosus, S. v. longicauda, S. v. alascensis* and *S. v. shumaginensis* see accounts of those subspecies. Paler and slightly larger than S. v. soperi. Larger than the subspecies from central Montana herein described as new. Smaller than *S. v. neomexicanus*. Averaging larger in all dimensions than *S. v. monticola* with which *obscurus* intergrades in northern New Mexico and northern Arizona. Larger than *S. v. vagrans* with more grayish rather than reddish fresh summer pelage and light gray rather than dark grayish-black fresh winter pelage.

Remarks.--Intergradation of *S. v. obscurus* with *S. v. setosus, S. v. longicauda, S. v. alascensis,* and the new subspecies from Montana takes place in the usual way with specimens from intermediate localities being intermediate in size and color. However the relationship of *S. v. obscurus* and *S. v. vagrans* (as the latter subspecies is defined in this study) is rather complicated. In southern British Columbia where the two subspecies come together a situation of remarkable complexity prevails. Series from some localities seem to represent intergrades between *obscurus* and *vagrans*; from other localities some specimens seem to be referable to one and some to the other subspecies; from other localities all specimens seem referable to one subspecies. A similar situation is seen in specimens from northeastern Washington, northern and central Idaho, and extreme western Montana. The region mentioned is one of extensive interfingering of life-zones. In southern British Columbia the main axes of the rivers, valleys and mountain ranges are north and south. Most of the valleys are in the Transition Life-zone; the forests are rather dry and of pine with more or less isolated hydrosere communities about streams and ponds. These hydrosere situations are the habitat of *Sorex vagrans*. Shrews from these situations are usually referable to *vagrans*. The high ridges and mountain ranges are usually in the Canadian Life-zone or higher and most of the shrews referable to *obscurus* come from such places. Marginal localities with regard to life-zone produce most of the populations which seem to represent intergrades between the two subspecies. Isolated areas of Canadian Life-zone, even though surrounded with Transition Life-zone, often harbor a population of *obscurus*, whereas the streams in the nearby dry valleys harbor populations of *vagrans*. Farther south in the Rocky Mountain chain, *obscurus* seemingly intergrades regularly with *vagrans*. This intergradation is seen in populations from several localities in Utah. There the lower elevations west of the Wasatch and Uinta mountains are inhabited by *S. v. vagrans*, the higher elevations by *obscurus* and where the ranges of the two abut intergrading populations occur. In these series of intergrades there are specimens which, using size as a subspecific criterion, would unhesitatingly be assigned, as individuals, to *obscurus*, and others would be assigned to *vagrans*, but these individuals represent extremes of a normally variable population. At Cuddy Mountain, Idaho, the two subspecies seemingly abut without intergradation; anyhow the available specimens from this locality are referable to one or the other subspecies and none is intermediate. The situation just described understandably has been the source of much anguish to students who sought to identify shrews from the Rocky Mountains. The reason for the relationship just described has been discussed at length in a previous section.

In the Rocky Mountains of Wyoming and Colorado the subspecies *S. v. obscurus* ranges almost uninterruptedly over relatively large areas, but southward in New Mexico and southwestward into Utah and Arizona, suitable boreal habitat becomes insular in nature and obscurus there is confined to the higher mountains. With one exception, once the shrew populations become 'insular' in this region they become smaller and show intergradation with *Sorex vagrans monticola*. The exception is the population in the Sacramento Mountains of southeastern New Mexico which is larger than *obscurus* and has been rightly recognized as a distinct subspecies, *neomexicanus*.

Almost without exception the range of typical *Sorex vagrans obscurus* is sympatric with that of *Sorex cinereus*, usually the subspecies *S. c. cinereus*. So close is this correspondence that the presence of *S. cinereus* comes near to being a useful aid in identifying *S. v. obscurus*. In areas where individuals of *obscurus* show intergradation with *vagrans*, *Sorex cinereus* is absent or rare. The implication is that as the species *S. vagrans* approaches the size of the species S. cinereus, competition between the two increases with resultant displacement of *cinereus*.

Specimens examined.--Total number, 982.

ALASKA: Wahoo Lake, 69deg 08' N, 146deg 58' W, 2350 ft., 2 KU; Chandler Lake, 68deg 12' N, 152deg 45' W, 2900 ft., 1 KU; Bettles, 1 KU, 5 BS; Alatna, 1 BS; Yukon River, 20 mi. above Circle, 1 BS; Tanana, 1 BS; Mountains near Eagle, 18 BS; Richardson, 8 BS; head of Toklat River, 11 BS; Savage River, 8 BS.

YUKON: MacMillan Pass, Mile 282, Canol Road, 1 NMC; MacMillan River, Mile 249, Canol Road, 1 NMC; S. fork MacMillan River, Mile 249, Canol Road, 2 NMC; Sheldon Lake, Mile 222, Canol Road, 5 NMC; Rose River, Mile 95, Canol Road, 1 NMC; McIntyre Creek, 3 mi. NW Whitehorse, 2250 ft. 1 KU; Nisutlin River, Mile 40, Canol Road, 6 NMC; SW end Dezadeash Lake, 2 KU; 3 mi. E and 11/2 mi. S Dalton Post, 2500 ft., 1 KU.

MACKENZIE: Nahanni River Mtns., Mackenzie River, 1 BS; Fort Simpson, 3 BS; Fort Resolution, Mission Island, 1 BS.

BRITISH COLUMBIA: W. side Mt. Glave, 4000 ft., 14 mi. S and 2 mi. E Kelsall Lake, 1 KU; Stonehouse Creek, 51/2 mi. W jct. Stonehouse Creek and Kelsall River, 4 KU; Bennett City, 6 BS; Wilson Creek, Atlin, 1 PMBC; McDame Post, Dease River, 6 BS; McDame Creek, 3 BS; Hot Springs, 3 mi. WNW jct. Trout River and Liard River, 1 KU; NW side Muncho Lake, 1 KU; Little Tahtlan River, 1 AMNH; Junction (4 mi. N Telegraph Creek), 7 BS; Raspberry Creek, 16 AMNH; Klappan River Valley, 1 BS; Chapa-atan River, 4 BS; Fort Grahame, 3 BS; Kispiox Valley, 23 mi. N Hazleton, 1 BS; Bear Lake, site of Fort Connully, 2 BS; Tetana Lake, 1 PMBC; Hudson Hope, 2 BS; Charlie Lake, 3 PMBC; Babine Mts., 6 mi. N Babine Trail, 5200 ft., 1 BS; Big Salmon River (S branch near Canyon), 1 BS; Ootsa Lake, 2 PMBC; Indianpoint Lake, 4 PMBC; Barkerville, 7 BS; Yellowhead Lake, 2 NMC, 1 PMBC; N. fork Moose River, 1 BS; Moose

Lake, 2 BS; Moose Pass, 1 BS; Glacier, 7 AMNH, 12 BS; Golden, 1 BS; Field, 2 BS; Caribou Lake, near Kamloops, 2 BS; Sicamous, 1 BS; Monashee Pass, 4 PMBC; Paradise Mine, 3 PMBC; Level Mtn., 4 AMNH; 6 mi. S Nelson, 6 BS; Morrissey, 5 NMC; Wall Lake, 1 BS.

ALBERTA: Hays Camp, Slave River, Wood Buffalo Park, 1 NMC; Kinuso, Assineau River, 1920 ft., 2 KU; Athabaska River, 30 mi. above Athabaska Landing, 8 BS; Smokey Valley, 50 mi. N Jasper House, 1 BS; Sulfur Prairie, Grande Cache River, 3 BS; Stoney River, 35 mi. N Jasper House, 1 BS; Moose Mtn., 1 NMC; Rodent Valley, 25 mi. W Henry House, 1 BS; Henry House, 3 BS; Jasper, 2 NMC; Shovel Pass, 4 NMC; mouth of Cavell Creek, Jasper Park, 1 NMC; 11 mi. S Henry House, 2 BS; 15 mi. S Henry House, 1 BS; Red Deer River, 1 AMNH; 27 mi. W Banff, 3 NMC; 12 mi. WNW Banff, 4500 ft., 1 NMC; N. Fork Saskatchewan River, 5000 ft., 1 NMC; Cypress Hills, 1 NMC; Waterton Lakes Park, 53 NMC.

SASKATCHEWAN: Cypress Hills, 21 NMC.

WASHINGTON: *Okanogan Co.*: Pasayten River, 1 BS; Bauerman Ridge, 1 BS; Conconully, 2 BS. *Pend Oreille Co.*: 2 mi. N Gypsy Meadows, 2 WSC; Round Top Mtn., 1 WSC; head Pass Creek, 1. *Chelan Co.*: Stehekin, 4 BS; head Lake Chelan, 4 BS; Wenatchee, 1 BS. *Kittitas Co.*: Easton, 10 BS.

IDAHO: *Boundary Co.*: Cabinet Mtns., E Priest Lake, 2 BS. *Adams Co.*: 1/2 mi. E Black Lake, 1 KU; 1 mi. N Bear Creek R. S., SW slope Smith Mtn., 2 KU. *Washington Co.*: 1 mi. NE Heath, SW slope Cuddy Mtn., 4000 ft., 4 KU. *Lemhi Co.*: 10 mi. SSW Leadore (type locality), 4 BS; 5 FC. *Fremont Co.*: 7 mi. W West Yellowstone, 4 KU. *Custer Co.*: head Pahsimeroi River, Pahsimeroi Mtns., 1 BS. *Blaine Co.*: Perkins Lake, 1 KU. *Bear Lake Co.*:--*Caribou Co.* line: Preuss Mts., 1 BS.

MONTANA: *Glacier Co.*: Sherburne Lake, 3 UM; 21/2 mi. W and 11/2 mi. S Babb, 1 KU; St. Mary's, 6 UM; St. Mary Lakes, 9 BS; Fish Creek, 2 BS; Gunsight Lake, 2 BS. *Flathead Co.*: Nyack, 3 UM, 1 BS; 1 mi. W and 2 mi. S Summit, 1 KU. *Ravalli Co.*: 8 mi. NE Stevensville, 3 BS; Sula, 1 BS. *Meagher Co.*: Big Belt Mtns., Camas Creek, 4 mi. S Fort Logan, 7 BS. *Gallatin Co.*: West Gallatin River, 4 BS. *Park Co.*: Emmigrant Gulch, 3 mi. SE Chico, 2 BS; Beartooth Mtns., 2 BS; *Carbon Co.*: Pryor Mtns., 2 BS.

WYOMING: *Yellowstone Nat'l Park*: Mammoth Hot Springs, 11 BS; Tower Falls, 1 BS; Astringent Creek, 1 BS; Flat Mtn., 1 BS; Yellowstone Park, 1 UM. *Park Co.*: Beartooth Lake, 15 BS; SW slope Whirlwind Peak, 1 KU; Pahaska Tepee, 6300 ft., 8 BS; Pahaska, mouth Grinnell Creek, 15 BS; Pahaska, Grinnell Creek, 7000-7500 ft., 18 BS; 25 mi. S and 28 mi. W Cody, 1 KU; Valley, Absaroka Mts., 14 BS; Needle Mtn., 2 BS. *Big Horn Co.*: 28 mi. E Lovell, 9000 ft., 12 KU; head Trapper's Creek, W slope

Bighorn Mtns., 7 BS; 171/2 mi. E and 41/2 mi. S Shell, 1 KU. *Teton Co.*: Two Ocean Lake, 6 FC; Emma Matilda Lake, 2 BS; 1 mi. N Moran, 1 FC; 21/2 mi. E and 1/4 mi. N Moran, 6230 ft., 2 KU; Moran, 7 FC, 1 KU; 21/2 mi. E Moran, 6220 ft., 1 KU; 1 mi. S Moran, 1 FC; 33/4 mi. E and 1 mi. S Moran, 9 KU; 7 mi. S Moran, 3 FC; Timbered Island, 14 mi. N Moose, 6750 ft., 3 KU; Bar BC Ranch, 21/2 mi. NE Moose, 6500 ft., 1 KU; Beaver Dick Lake, 1 UM; Teton Mtns., Moose Creek, 6800 ft., 9 BS; Teton Mtns., S Moose Creek, 10,000 ft., 3 BS; Teton Pass, above Fish Creek, 7200 ft., 15 BS; Whetstone Creek, 4 UM; Flat Creek-Gravel Creek Divide, 2 UM; Flat Creek-Granite Creek Divide, 1 UM; Jackson, 3 KU, 2 UM. *Fremont Co.*: Togwotee Pass, 5 FC; Jackey's Creek, 3 mi. S Dubois, 1 BS; Milford, 5400 ft., 2 KU; Mosquito Park R. S. 171/2 mi. W and 21/2 mi. N Lander, 1 KU; 17 mi. S and 61/2 mi. W Lander, 9300 ft., 1 KU; Mocassin Lake, 19 mi. W and 4 mi. N Lander, 10,000 ft., 1 KU; 231/2 mi. S and 5 mi. W Lander, 8600 ft., 1 KU; Green Mts., 8 mi. E Rongis, 8000 ft., 4 BS. *Washakie Co.*: 9 mi. E and 5 mi. N Tensleep, 7400 ft., 2 KU; 9 mi. E and 4 mi. N Tensleep, 7000 ft., 2 KU. *Lincoln Co.*: Salt River Mtns., 10 mi. SE Afton, 5 BS; Labarge Creek, 9000 ft., 1 BS. *Sublette Co.*: 31 mi. N Pinedale, 8025 ft., 3 KU; Surveyor's Park, 12 mi. NE Pinedale, 8000 ft., 2 BS; N. side Half Moon Lake, 7900 ft., 1 KU; 21/2 mi. NE Pinedale, 7500 ft., 2 KU. *Natrona Co.*: Rattlesnake Mts., 7000-7500 ft., 18 BS; Casper Mts., 7 mi. S Casper, 6 BS. *Converse Co.*: 211/2 mi. S and 241/2 mi. W Douglas, 7600 ft., 7 KU; 22 mi. S and 241/2 mi. W Douglas, 7600 ft., 4 KU; 221/2 mi. S and 241/2 mi. W Douglas, 7600 ft., 2 KU. *Uinta Co.*: 1 mi. N Fort Bridger, 6650 ft., 1 KU; Fort Bridger, 3 KU; Evanston, 1 BS; 9 mi. S Robertson, 8000 ft., 6 KU; 9 mi. S and 21/2 mi. E Robertson, 8600 ft., 1 KU; 10 mi. S and 1 mi. W Robertson, 8700 ft., 3 KU; 101/2 mi. S and 2 mi. E Robertson, 8900 ft., 1 KU; 13 mi. S and 1 mi. E Robertson, 9000 ft., 1 KU; 13 mi. S and 2 mi E Robertson, 9200 ft., 1 KU. *Carbon Co.*: Ferris Mts., 7800 to 8500 ft., 13 BS; Shirley Mts., 7600 ft., 7 BS; Bridget's Pass, 18 mi. SW Rawlins, 7500 ft., 2 KU; 10 mi. N and 12 mi. E Encampment, 7200 ft., 1 KU; 10 mi. N and 14 mi. E Encampment, 8000 ft., 6 KU; 91/2 mi. N and 111/2 mi. E Encampment, 7200 ft., 2 KU; 9 mi. N and 3 mi. E Encampment, 6500 ft., 1 KU; 9 mi. N and 8 mi. E Encampment, 7000 ft., 1 KU; 8 mi. N and 14 mi. E Encampment, 8400 ft., 3 KU; 8 mi. N and 141/2 mi. E Encampment, 8100 ft., 2 KU; 8 mi. N and 16 mi. E Encampment, 4 KU; 8 mi. N and 211/2 mi. E Encampment, 9400 ft., 2 KU; S. base Bridger's Peak, 8800 ft., Sierra Madre Mts., 3 BS; 8 mi. N and 191/2 mi. E Savery, 8800 ft., 2 KU; 7 mi. N and 17 mi. E Savery, 8300 ft., 1 KU; 61/2 mi. N and 16 mi. E Savery, 8300 ft., 1 KU; 6 mi. N and 15 mi. E Savery, 8500 ft., 1 KU; 5 mi. N and

101/2 mi. E Savery, 8000 ft., 2 KU; 14 mi. E and 6 mi. S Saratoga, 8800 ft., 1 KU. *Albany Co.*: Springhill, 12 mi. N Laramie Peak, 6300 ft., 10 BS; Laramie Peak, N. slope, 8000 to 8800 ft., 7 BS; Bear Creek, 3 mi. SW Laramie Peak, 7500 ft., 6 BS; 21/2 mi. ESE Brown's Peak, 10,500 ft., 2 KU; 3 mi. ESE Brown's Peak, 10,000 ft., 1 KU; 27 mi. N and 5 mi. E Laramie, 6960 ft., 2 KU; 1 mi. SSE Pole Mtn., 8350 ft., 3 KU; 2 mi. SW Pole Mtn., 3 KU; 3 mi. S Pole Mtn., 8100 ft., 2 KU; 83/4, mi. E and 61/2 mi. S Laramie, 8200 ft., 2 KU; Woods P. O., 1 BS. *Laramie Co.*: 5 mi. W and 1 mi. N Horse Creek P. O., 7200 ft., 2 KU.

UTAH: *Weber Co.*: Mt. Willard, Weber-Box Elder Co. line, 9768 ft., 2 UU. *Salt Lake Co.*: Butterfield Canyon, 7000 ft., 1 UU; Brighton, Silver Lake P. O., 8700 ft., 2 UU; Brighton, Silver Lake P. O., 8750 ft., 8 UU; Brighton, Silver Lake P. O., 9000 ft., 2 UU; Brighton, Silver Lake P. O., 9500 ft., 1 UU. *Summit Co.*: Jct. Bear River and East Fork, 2 CM; Smith and Morehouse Canyon, 7000 ft., 1 UU; Mirror Lake, 10,000 ft., 1 UU. *Daggett Co.*: Jct. Deep and Carter creeks, 7900 ft., 1 UU. *Utah Co.*: Nebo Mtn., 1 mi. E Payson Lake, 8300 ft., 1 UU; Nebo Mts., 12 mi. SE Payson Lake, 1 UU. *Wasatch Co.*: Current Creek, Uinta Mts., 1 BS; Wasatch Mts., 1 BS. *Uintah Co.*: Paradise Park, 21 mi. W and 15 mi. N Vernal, Uinta Mts., 10,050 ft., 2 CM, 3 KU; Paradise Park, Uinta Mts., 10,100 ft., 6 UU. *Sanpete Co.*: Manti, 3 BS. *Sevier Co.*: 7 mi. Creek, 20 mi. SE Salina, 5 CM; Fish Lake Plateau, 2 BS. *Emery Co.*: Lake Creek, 11 mi. E Mt. Pleasant, 4 CM. *Grand Co.*: Warner R. S., La Sal Mts., 9750 ft., 2 UU; La Sal Mts., 11,000 ft., 1 BS. *Beaver Co.*: Puffer Lake, Beaver Mts., 2 BS. *Wayne Co.*: Elkhorn G. S., Fish Lake Plateau, 14 mi. N Torrey, 9400 ft., 3 UU. *Garfield Co.*: Wildcat R. S., Boulder Mtn., 8700 ft., 6 UU; 18 mi. N Escalante, 9500 ft., 1 UU. *Washington Co.*: Pine Valley Mts., 7 BS. *San Juan Co.*: Geyer Pass, 18 mi. SSE Moab, 3 CM; Cooley, 8 mi. W Monticello, 3 CM.

COLORADO: *Larimer Co.*: Poudre River, 1 KU. *Rio Blanco Co.*: 91/2 mi. SW Pagoda Peak, 2 KU. *Grand Co.*: Arapaho Pass, Rabbit Ears Mts., 2 BS. *Boulder Co.*: Willow Park, Rocky Mtn. Nat'l Park, 8 UM; Longs Peak, 1 BS; 3/4 mi. N and 2 mi. W Allenspark, 8400 ft., 5 KU; Ward, 9500 ft., 1 BS; Buchanan Pass, 1 BS; 3 mi. S Ward, 1 KU; 7 mi. NW Nederland, 1 KU; 5 mi. W Boulder, 3 BS; Boulder, 3 BS, 1 ChM; Nederland, 6 BS, 4 ChM; Eldora, 1 BS. *Garfield Co.*: Baxter Pass, 8500 ft., 2 BS. *Eagle Co.*: Gores Range, 1 BS. *Gilpen Co.*: Black Hawk, 1 BS. *Lake Co.*: 3 mi. W Twin Lakes, 2 KU; 12 mi. S and 1 mi. W Leadville, 1 KU. *Gunnison Co.*: 2 mi. W Gothic, 2 FC; Copper Lake, 2 FC; Gothic, 1 FC. *Chaffee Co.*: St. Elmo, 10,100 ft., 2 BS; E side Monarch Pass, 7 mi. W Salida, 2 ChM. *Teller Co.*: Glen Core, Pikes Peak, 2 UM. *El Paso Co.*: Hunters Creek, a tributary of

Bear Creek, 7250-7400 ft., 1 AMNH. *Montrose Co.*: Uncomphagre Plateau, 8500 ft., 3 BS. *Saguache Co.*: 3 mi. N and 16 mi. W Saguache, 8500 ft., 2 KU; Cochetopa Pass, 10,000 ft., 4 KU; Monshower Meadows, 27 mi. W Saguache, 2 BS. *San Juan Co.*: Silverton, 4 BS. *Mineral Co.*: 23 mi. S and 11 mi. E Creede, 1 KU. *Costilla Co.*: Fort Garland, 2 BS. *Huerfano Co.*: 5 mi. S and 1 mi. W Cuchara Camps, 8 KU.

NEW MEXICO: *Taos Co.*: 3 mi. N Red River, 2 BS; Taos, 1 BS. *Colfax Co.*: 1 mi. S and 2 mi. E Eagle Nest, 8100 ft., 2 KU. *Sandoval Co.*: Jemez Mts., 3 BS. *Santa Fe Co.*: Hyde Park, 5 mi. NE Santa Fe, 2 HC; Santa Fe Field Station, 1 HC; Santa Fe Ski Basin, 1 KU; Pecos Baldy, 4 BS. *Torrance Co.*: Manzano Mts., 2 BS.

Marginal Records.--ALASKA: Chandler Lake, 68deg 12' N, 152deg 45' W; Yukon River, 20 mi. above Circle; Mountains near Eagle. MACKENZIE: Nahanni River Mts.; Fort Simpson; Fort Resolution, Mission Island. ALBERTA: Wood Buffalo Park; Athabaska River, 30 mi. above Athabaska Landing. SASKATCHEWAN: Cypress Hills. MONTANA: St. Mary; 4 mi. S Fort Logan; Pryor Mts. WYOMING: 1 mi. W and 1 mi. S Buffalo, 27424 KU; Springhill, 12 mi. N Laramie Peak; 5 mi. W and 1 mi. N Horse Creek PO. COLORADO: Boulder; Hunters Creek; 5 mi. S and 1 mi. W Cuchara Camps. NEW MEXICO: 3 mi. N Red River, 10,700 ft.; Pecos Baldy; Manzano Mts.; Jemez Mts. COLORADO: Navajo River (Jackson, 1928:120); Silverton. UTAH: La Sal Mts., 11,000 ft. COLORADO: Baxter Pass. UTAH: junction Trout and Ashley Creeks, 9700 ft. (Durrant, 1952:35); Mirror Lake, 10,000 ft.; Mt. Baldy R. S. (Durrant, 1952:53); Wildcat R. S.; Pine Valley Mts.; Puffer Lake; Butterfield Canyon. IDAHO: Preuss Mts.; 4 mi. S Trude (Davis, 1939:104); head Pahsimeroi River, Pahsimeroi Mts.; Perkins Lake; 1 mi. NE Heath; 1/2 mi. E Black Lake. MONTANA: Sula; 8 mi. NE Stevensville. WASHINGTON: head Pass Creek; Conconully; Wenatchee; Easton; Stehekin; Pasayten River. BRITISH COLUMBIA: Second Summit, Skagit River, 5000 ft., (Jackson, 1928:120); Babine Mts., 6 mi. N Babine Trail, 5200 ft.; Hazleton (Jackson, 1928:120); 23 mi. N Hazleton; Flood Glacier, Stikine River (Jackson, 1928:120); Cheonee Mts. (*ibid.*); Level Mtn.; west side Mt. Glave, 4000 ft., 14 mi. S and 2 mi. E Kelsall Lake. ALASKA: head Toklat River; Tanana; Alatna; Bettles.

Sorex vagrans soperi Anderson and Rand

Sorex obscurus soperi Anderson and Rand, Canadian Field-Nat., 59:47, October 16, 1945.

Type.--Adult male, skin and skull; No. 18249, Nat. Mus. Canada; obtained on September 21, 1940, by J. Dewey Soper, from 21/2 mi. NW Lake Audy, Riding Mtn. Nat'l Park, Manitoba.

Range.--Southwestern Manitoba to central Saskatchewan.

Diagnosis.--Size medium to small for the species; measurements of type and two topotypes are: total length, 107, 108, 117; tail, 45, 45, 45; hind foot, 12.1, 12.3, 12.5. Color dark brownish or fuscous in summer pelage; winter pelage unknown.

Comparison.--Resembles *S. v. obscurus* in size; darker than *obscurus* in summer pelage; cranium slightly higher and top more nearly flat; larger and darker in summer pelage than the new subspecies from central Montana.

Remarks.--In their description of this subspecies Anderson and Rand pointed out that specimens from the type locality and from central Saskatchewan represent the dark extreme in a color cline which begins in south-central British Columbia with "pale, brownish-tinged animals." These authors referred shrews from Cypress Hills, southwestern Saskatchewan and southeastern Alberta to *S. o. soperi*, although they noted that these specimens, taken by themselves, are not strikingly different from *S. o. obscurus* from the Rocky Mountains. The specimens from the Cypress Hills were included in *soperi* because the authors felt that the subspecific boundary should be drawn "where specimens average about half way between the extremes (of the cline) in characters."

It is true, as Anderson and Rand say, that the shrews from Cypress Hills are hardly separable from those from, say, Waterton Lakes Park. The specimens from the Cypress Hills are noticeably different from specimens from the Okanagan area, but some of the latter, in my opinion may represent intergrades between *S. v. obscurus* and the more reddish *S. v. vagrans* and are not, at any rate, typical *obscurus*. In view of the similarity of shrews from Cypress Hills to typical *S. v. obscurus* and since the Cypress Hills are much nearer to the range of *S. v. obscurus* than to the record-stations of occurrence in central Saskatchewan and Manitoba, I have chosen to restrict the name *soperi* to shrews from these latter two localities. Seemingly *S. vagrans* is absent from the plains separating the Cypress Hills from the Rocky Mountains and from Riding Mountain National Park.

Specimens examined.--none.

Marginal records.--SASKATCHEWAN: Prince Albert National Park, 1700 ft. (Anderson and Rand, 1945:48). MANITOBA: Riding Mountain National Park, 21/2 mi. NW Audy Lake (ibid.).

Sorex vagrans longiquus new subspecies

Type.--First year male, skin and skull; No. 87332, Univ. Michigan Mus. Zool.; obtained on July 21, 1942, by Emmet T. Hooper from 25 mi. ESE Big Sandy, Eagle Creek, Chouteau Co., Montana, original no. 2184.

Range.--Central Montana; marginal localities are: Bearpaw Mts., Zortman, Big Snowy Mts., Buffalo, Little Belt Mts.

Diagnosis.--Size small for the species; measurements of three topotypes are: total length, 101, 105, 108; tail, 39, 40, 42; hind foot, 11.5, 11.5, 12. Color pale; summer pelage: back near (17 *k*) Olive Brown but hairs of dorsum with a pale, buffy band proximal from the tips which imparts a pale over-all appearance; flanks near Wood Brown; underparts Pale Smoke Gray, usually not with a buffy wash; color of underparts often extending along margin of upper lip. Skull small for species; rostrum relatively broad and heavy; relatively broad interorbitally.

Comparisons.--From *S. v. obscurus, S. v. longiquus* differs as follows: size smaller; skull smaller in all dimensions although similar in proportion. From *S. v. soperi, S. v. longiquus* differs in: size smaller; color paler in summer pelage. From *S. v. vagrans, S. v. longiquus* differs in: color paler in summer pelage, less brownish; color of venter extending higher on flanks; venter Pale Smoke Gray, rarely tinged with buffy rather than usually tinged with buffy. From *S. v. monticola, S. v. longiquus* differs in: summer pelage slightly paler, venter Pale Smoke Gray rather than suffused with buffy.

Remarks.--The subspecies *longiquus* is obviously derived from the neighboring *S. v. obscurus* and differs from it mainly in size. Some specimens of obscurus from western Montana show evidences of intergradation with *S. v. vagrans* in possessing a somewhat buffy belly and these are thus more strikingly different from *longiquus* than are other specimens of *obscurus*. Many specimens of *obscurus* from the eastern slope of the Lewis and Clark Range in Montana show the tricolored pattern seen in many specimens of *longiquus*. The smallest individuals of longiquus are found on the Big Snowy Mountains. Intergradation with *obscurus* is seen in specimens here referred to *S. v. obscurus* from the Big Belt Mountains.

Specimens examined.--Total number, 45. MONTANA: *Hill Co.*: Bearpaw Mts., 5 UM, 2 BS. *Phillips Co.*: Zortman, 1 BS. *Chouteau Co.*: type locality,

3 UM; Highwood Mts., 13 BS. *Cascade Co.*: Neihart, Little Belt Mts., 1 BS. *Judith Basin Co.*: 3 mi. W Geyser, 4100 ft., 1 KU; Otter Creek, 10 mi. SW Geyser, 1 BS; Dry Wolf Creek, 20 mi. SW Stanford, 1 BS. Buffalo, 13 mi. W Buffalo Canyon, 2 BS. *Fergus Co.*: Moccasin Mts., 15 mi. NW Hilger, 3 BS; Judith Mts., 17 mi. NE Lewiston, 1 BS; 15 mi. S Heath, N. fork Flat Willow Creek, Big Snowy Mts., 1 BS; Timber Creek, Big Snowy Mts., 1 BS; Crystal Lake, 6000 ft., Big Snowy Mts., 2 UM; Rocky Creek, 5600 ft., Big Snowy Mts., 3 UM; Big Snowy Mts., 3 BS. *Meagher Co.*: Sheep Creek, 16 mi. N White Sulphur Springs, Little Belt Mts., 1 BS.

Marginal records.--MONTANA: Bearpaw Mts.; Zortman; Big Snowy Mts.; 16 mi. N White Sulphur Springs; Highwood Mts.

Sorex vagrans neomexicanus Bailey

Sorex obscurus neomexicanus Bailey, Proc. Biol. Soc. Washington, 26:133, May 21, 1913.

Type.--Adult male, skin and skull; No. 100440, U. S. Biol. Surv. Coll.; obtained on May 29, 1900, by Vernon Bailey, from Cloudcroft, 9000 ft., Otero Co., New Mexico.

Range.--Sacramento and Capitan Mountains of New Mexico.

Diagnosis.--Size medium for the species; average and extreme measurements of 4 topotypes are: total length, 105.2 (103-107); tail, 41.0 (39-42); hind foot, 13.1 (12.5-14). Color near Olive Brown in summer; winter pelage unknown; skull large and relatively broad; teeth relatively large.

Comparisons.--Skull larger than that of *S. v. obscurus* and relatively somewhat broader; much larger in all cranial dimensions than *S. v. monticola*.

Remarks.--*S. v. neomexicanus* is a well-marked subspecies seemingly limited to the mountains of southeastern New Mexico. It is the only species of *Sorex* thus far recorded from that area.

Specimens examined.--Total number, 12. NEW MEXICO: *Otero Co.*: SW slope Capitan Mts., 2 BS; 10 mi. NE Cloudcroft, 2 BS; Cloudcroft, 7 BS, 1 UM.

Marginal records.--NEW MEXICO: SW slope Capitan Mts.; 10 mi. NE Cloudcroft; type locality.

Sorex vagrans monticola Merriam

Sorex monticolus Merriam, N. Amer. Fauna, 3:43, September 11, 1890.

Sorex vagrans monticola, Merriam, N. Amer. Fauna, 10:69, December 31, 1895.

Sorex melanogenys Hall, Jour. Mamm., 13:260, August 9, 1932, type from Marijilda Canyon, 8600 ft., Graham Mts. [= Pinaleno Mts.] Graham Co., Arizona.

Type.--Adult male, skin and skull; No. 17599/24535, U. S. Biol. Surv. Coll.; obtained on August 28, 1899, by C. Hart Merriam and Vernon Bailey from San Francisco Mtn., 11,500 ft., Coconino Co., Arizona.

Range.--Mountains of western New Mexico, eastern Arizona, and the northern Sierra Madre Occidental of Mexico.

Diagnosis.--Size small for the species; average and extreme measurements of 12 specimens from the White Mountains, Arizona, are: total length, 104.3 (98-112); tail, 41.2 (37-45); hind foot, 12.0 (11-13). Summer pelage between (15 *m*) Proutts Brown and (15 *m*) Bister, venter tinged with (15 *f*) Pale Ochraceous Buff; winter pelage near (17 *k*) Olive Brown; skull relatively broad.

Comparisons.--For comparisons with *S. v. obscurus* and *S. v. neomexicanus* see accounts of those subspecies. Skull slightly larger and relatively broader than that of *S. v. orizabae,* and color slightly paler. Differs from *S. v. vagrans* in: winter pelage grayish (near 17 *k* Olive Brown) rather than blackish (17 *k* or 17 *m* Chaetura Drab or Chaetura Black); summer pelage slightly grayer; skull relatively slightly broader rostrally and interorbitally.

Remarks.--S. v. monticola intergrades gradually with *S. v. obscurus* to the north and east; indeed the type locality is actually in this area of intergradation. So far as I know, *monticola* is not in reproductive continuity with any other subspecies of *Sorex vagrans*. Specimens from southeastern Arizona are the smallest and seem to be the most "typical" in the sense that they are most different from *S. v. obscurus*. Some specimens from the whole length of the Rocky Mountain chain in the United States have for years been referred to *monticola*. Some of these, as I have pointed out, belong to *S. v. longiquus,* and others are intergrades between *S. v. obscurus* and *S. v. vagrans*. Since *vagrans* and *monticola* resemble one another somewhat, and since topotypes of *S. v. monticola* actually show the influence of intergradation with *obscurus,* it is easy to understand how intergrades between *obscurus* and *vagrans* could have been assigned to *monticola*.

Throughout most of its range, *S. v. monticola* is the only *Sorex* present. In some places *monticola* may occur with *S. nanus* or *S. merriami*. *S. v. monticola* occurs with the water shrew in southeastern Arizona.

Specimens examined.--Total number, 80.

ARIZONA: *Coconino Co.*: San Francisco Mtn., 3 BS, 6 CMNH. *Apache Co.*: Spruce Creek, Tunitcha Mts., 7 BS; Springerville, 1 BS; North Fork White River, White Mts., 12 SD; White River, Horseshoe Cienega, 8300 ft., White Mts., 5 BS; Mt. Thomas, 9500 to 11,000 ft., White Mts., 12 BS; Little Colorado River, White Mts., 4 BS; near head Burro Creek, 9000 ft., White Mts., 1 BS. *Graham Co.*: Graham Mts., 9200 ft., 2 BS. *Greenlee Co.*: Prieto Plateau, 9000 ft., S. end Blue Range, 1 BS. *Pima Co.*: Summerhaven, 7500 ft., Santa Catalina Mts., 3 BS, 1 SD. *Cochise Co.*: Fly Park, Chiricahua Mts., 4 BS; Rustler Park, Chiricahua Mts., 1 SD; Long Park, Chiricahua Mts., 1 UM; Huachuca Mts., 1 BS. *Santa Cruz Co.*: Stone Cabin Canyon, 8500 ft., Santa Rita Mts., 1 BS.

NEW MEXICO: *San Juan Co.*: Chusca Mts., 1 BS. *Catron Co.*: Mogollon Mts., 3 BS; 10 mi. E Mogollon, 1 KU. *Socorro Co.*: Copper Canyon, Magdalena Mts., 3 BS. *Sierra Co.*: Mimbres Mts., near Kingston, 1 BS.

CHIHUAHUA: Sierra Madre, near Guadalupe y Calvo, 5 BS.

Marginal records.--ARIZONA: Tunitcha Mts. NEW MEXICO: Chusca Mts.; Copper Canyon, Magdalena Mts.; Mimbres Mts., near Kingston. CHIHUAHUA: Guadalupe y Calvo. ARIZONA: Huachuca Mts.; Santa Catalina Mts.; White River, Horseshoe Cienega, 8300 ft., White Mts.; San Francisco Mtn.

Sorex vagrans orizabae Merriam

Sorex orizabae Merriam, N. Amer. Fauna, 10:71, December 31, 1895.

Sorex vagrans orizabae, Jackson, N. Amer. Fauna, 51:113, July 24, 1928.

Type.--Adult female, skin and skull; No. 53633, U. S. Biol. Surv. Coll.; obtained on April 24, 1893, by E. W. Nelson from W slope of Mt. Orizaba, 9,500 ft., Puebla.

Range.--Transverse volcanic belt of mountains at the southern end of the Mexican Plateau.

Diagnosis.--Size small for the species; measurements of 3 specimens from Volcan Toluca, Mexico, are: total length, 98, 100, 108; tail, 35, 39, 40; hind foot, 13, 13, 14. Summer pelage Mummy Brown tending toward Olive Brown; Fuscous to Fuscous-Black in winter; skull and teeth relatively narrow.

Comparisons.--For comparison with *S. v. monticola* see account of that subspecies.

Remarks.--The range of *S. v. orizabae* probably is not now in contact with that of any other subspecies of *S. vagrans*, although judging by the slight degree of difference between *orizabae* and *monticola* the separation between the two has not been of great duration.

Sorex vagrans orizabae occurs with *S. saussurei saussurei* throughout the transverse volcanic belt.

Specimens examined.--Total number, 23.

MICHOACAN: Patamban, 1 BS; Nahuatzin, 3 BS; Mt. Tancitaro, 4 BS.

MEXICO: Salazar, 2 BS, 1 KU; Volcan de Toluca, 3 BS.

TLAXCALA: Mt. Malinche, 2 BS.

PUEBLA: Mt. Orizaba, 6 BS.

VERACRUZ: Cofre de Perote, 1 BS.

Marginal records.--MICHOACAN: *Patamban*. VERACRUZ: Cofre de Perote. PUEBLA: *Mt. Orizaba*. MICHOACAN: Mt. Tancitaro.

Sorex vagrans vagrans Baird

Sorex vagrans Baird, Rep't Pacific R. R. Survey 8: pt. 1, Mammals, p. 15, July 14, 1858.

Sorex suckleyi Baird, Rep't Pacific R. R. Survey 8: pt. 1, Mammals, p. 18, July 14, 1858, type from Steilacoom, Pierce Co., Washington.

Sorex dobsoni Merriam, N. Amer. Fauna, 5:33, July 30, 1891, type from Alturas or Sawtooth Lake, altitude about 7200 ft., E base Sawtooth Mts., Blaine Co., Idaho.

Sorex amoenus Merriam, N. Amer. Fauna, 10:69, December 31, 1895, type from near Mammoth, 8000 ft., head Owens River, E slope Sierra Nevada, Mono Co., California.

Sorex nevadensis Merriam, N. Amer. Fauna, 10:71, December 31, 1895, type from Reese River, 6000 ft., Nye-Lander Co. line, Nevada.

Sorex shastensis Merriam, N. Amer. Fauna, 16:87, October 28, 1899, type from Wagon Camp, Mt. Shasta, 5700 ft., Siskiyou Co., California.

Type.--Adult male, alcoholic; No. 1675, U. S. Nat. Mus.; obtained at Shoalwater (Willapa) Bay, Pacific Co., Washington; received from J. G. Cooper, and entered in Museum catalog on October 23, 1856.

Range.--The Great Basin, and Columbia Plateau west across the mountains to the Pacific coast of northern California, Oregon, Washington and southwestern British Columbia.

Diagnosis.--Size small for the species; average and extreme measurements of 8 topotypes are: total length, 104.1 (99-109); tail, 43.3 (42-45); hind foot, 12.9 (12-14). Summer pelage ranging from (15 *k*)

Cinnamon Brown through (15 *m*) Proutt's Brown to (17 *m*) Mummy Brown. Winter pelage (13 *m*) Fuscous Black to (17 *m*) Chaetura Black.

Comparisons.--For comparison with *S. v. monticola* see account of that subspecies. Differs from *S. v. halicoetes* in relatively narrower and more attenuate rostrum and in less brownish underparts in winter pelage; smaller and more brownish (less grayish) than *Sorex vagrans* from the southern Sierra Nevada.

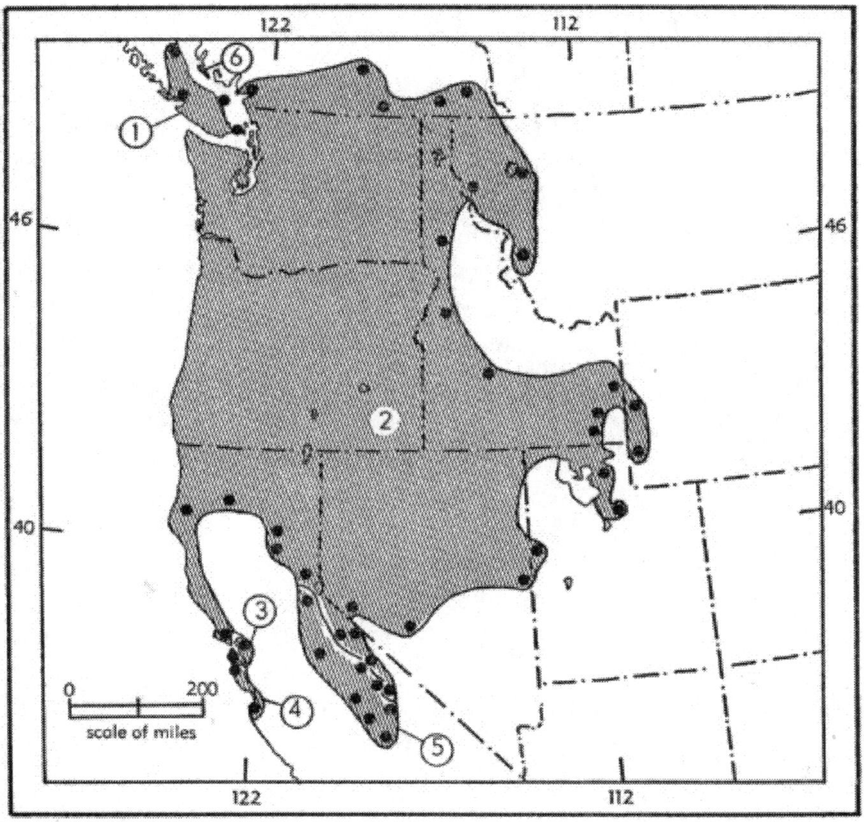

FIG. 18. Probable geographic ranges of *Sorex vagrans vagrans*, its derivative subspecies, and *S. v. mixtus*.

1. *S. v. vancouverensis*
2. *S. v. vagrans*
3. *S. v. halicoetes*

4. *S. v. paludivagus*
5. *S. v. obscuroides*
6. *S. v. mixtus*

Remarks.--Restriction of the range of *S. v. monticola* to Arizona and New Mexico leaves shrews that were formerly assigned to this subspecies from Utah, Idaho, Washington and southern British Columbia unassigned. Several names are available for consideration. The name *Sorex vagrans dobsoni* Merriam, 1891, type locality Alturas Lake, Blaine Co., Idaho, was once applied to small shrews from Idaho, Montana, Wyoming, and Utah, but was considered by Jackson to be synonymous with *S. v. monticola*. The name *Sorex vagrans amoenus* Merriam, 1895, type locality near Mammoth, Mono Co., California, has been applied to wandering shrews from western Nevada, northeastern California and southern Oregon. *Sorex vagrans nevadensis* Merriam, 1895, type locality Reese River on Nye-Lander Co. line, Nevada was considered by Hall (1946:119) to be synonymous with *S. v. amoenus*. Specimens of *Sorex vagrans* west of the Cascade Mountains have long been referred to the nominate subspecies which has its type locality at Willapa Bay, Pacific Co., Washington. Over so wide an area it is only to be expected that some geographic variation is to be found. Thus specimens from central Nevada average slightly paler in summer pelage than those from the Pacific Coast or from the foothills of the Rocky Mountains. In addition there are slight average differences in size from place to place. Topotypes of *S. v. vagrans*, however, show a fair degree of variability and some are nearly as pale as the paler Great Basin stocks. Furthermore topotypical individuals of *vagrans* can be lost in series of *S. v. amoenus,* although *amoenus* is shorter-tailed on the average. Specimens from the western foothills of the Rocky Mountains show an amazing series of relationships with the montane *S. v. obscurus*. In Utah, as previously pointed out, complete intergradation occurs. At 1 mi. N Heath, Washington Co., Idaho, the lowland and the highland forms approach each other within a short distance and still maintain a degree of distinctness, especially in size. In northwestern Montana intergradation is extensive (Clothier, 1950). In northeastern Washington distinctly separable populations occur within a few miles of one another. In southern British Columbia some populations are clearly intergrades while at 6 mi. S Yahk intergradation seemingly has not taken place. Where some intergradation has occurred the result often has been increased size of the lowland shrews, although they usually retain the reddish summer pelage rather than acquiring the more grayish pelage of *obscurus*. The name *dobsoni* was based upon shrews from a place where lowland and highland forms occur almost together with only a slight amount of intergradation. Examples of "*dobsoni*" may not with certainty be distinguished from typical *vagrans* except that they are, as Merriam (1895:68-69) points out, somewhat larger. Merriam (*loc. cit.*) further notes that *dobsoni* is "intermediate in size and cranial characters between *S. vagrans* and *obscurus*;" a statement which hits very close to the heart of the matter. I consider the name *dobsoni* to apply to intergrades. To attempt to apply the name to the highly variable populations of intergrades from British Columbia to southern Idaho seems inadvisable. I have examined the possibility of using the name *amoenus* for the animals from this region. The characters which set *amoenus* apart from *vagrans*, slightly shorter tail and slightly darker summer pelage, however, are not universally found in shrews from the Columbian Plateau and eastern Great Basin and furthermore these differences between *amoenus* and *vagrans* do not seem to me to be of great enough magnitude to warrant subspecific recognition of the former. Thus the name *S. v. vagrans* may apply to shrews in the region under consideration. The subspecies, as thus thought of, embraces several incipient subspecies, namely (1) the populations on the isolated mountain ranges of Nevada, (2) the coastal rain forest population and possibly (3) the population on the Columbian Plateau.

In western British Columbia, Washington, and Oregon no evidences of intergradation between *S. v. vagrans* and the races *setosus*, *permiliensis*, *bairdi*, *yaquinae*, or *pacificus* are seen. In this region *S. v. vagrans* occurs sympatrically with one or the other of these subspecies. Different degrees of differentiation thus obtain between the subspecies *vagrans* as here defined and the surrounding subspecies of *Sorex vagrans* to wit: complete intergradation and allopatry in Utah with *S. v. obscurus*; partial intergradation and partial sympatry with *S. v. obscurus* in the foothill region from Idaho to British Columbia; no intergradation and complete sympatry with all the other races of *Sorex vagrans* from the Cascades to the coast and south to San Francisco Bay. The relationship of *S. v. vagrans* to the wandering shrews of the high Sierra is discussed on page 58.

Throughout most of the Great Basin and Columbian Plateau *Sorex vagrans* is, with the exception of the rare *S. merriami* and *S. preblei*, the only small shrew. In the Cascades and in the coastal lowlands it is the only small shrew except for *S. cinereus* and *S. trigonirostris*, both extremely rare and local in this region. *S. vagrans* seemingly competes to a certain extent with the larger *S. trowbridgii* in western Washington and seems to be partially dominant to *trowbridgii*, at least in marshy habitats (Dalquest, 1941:171).

Specimens examined.--Total number, 1197.

BRITISH COLUMBIA: *Osoyoos District*: Okanagan, 20 PMBC; Okanagan Landing, 2 PMBC; Nahun Plateau, 2 PMBC. *Vancouver District*: Vancouver, 2 PMBC. *New Westminister District*: Port Moody, 16 BS; Westminster Jct., 4 AMNH; Langley, 1 BS; Vedder Crossing, 1 PMBC; Huntingdon, 69 NMC; Sumas, 16 BS; Cultus Lake, 1 NMC. *Similkameen District*: Princeton, 6 Mile Creek, 1 NMC. Hedley, Stirling Creek, 7 NMC; Fairview-Keremeos Summit, 5 NMC; Oliver, 1 NMC; Westbridge, 6 NMC; Osoyoos, 1 PMBC; Osoyoos-Bridesville Summit, 4 NMC; Cascade, 7 NMC. *Nelson District*: Kuskonook, 1 PMBC; Rossland, 14 NMC; Trail, 2 NMC; Creston, 4 PMBC, 4 NMC; near Creston, 7 NMC. *Cranbrook District*: Cranbrook, 5 BS; Yahk, 2 NMC; Yahk Camp 6, 2 NMC; Goatfell, 2 NMC. *Fernie District*: Newgate, 3 NMC.

WASHINGTON: *Whatcom Co.*: Blaine, 1 BS; Beaver Creek, 5 WSC; Glacier, 1 BS; Mt. Baker Lodge, 1 WSC; Lake Whatcom, 1 BS; Barron, 2 BS. *Okanogan Co.*: Sheep Mtn., 3 BS; E. end Bauerman Ridge, 1 BS; Oroville, 1 BS; Hidden Lakes, 1 BS; Loomis, 1 BS; Conconully, 1 BS; Twisp, 1 BS. *Ferry Co.*: 5 mi. W Curlew, 2 BS. *Stevens Co.*: Marcus, 1 BS. *Pend Oreille Co.*: Canyon, 1 WSC; Metaline, 2 BS; Sullivan Lake, 1 BS. *San Juan Co.*: East Sound, Orcas Island, 3 BS; Friday Harbor, San Juan Island, 1 BS; San Juan Park, 2 WSC; Blakely Island, 1 KU; Richardson, 6 BS. *Skagit Co.*: Cypress Island, 1 KU; Hamilton, 1 BS; Sauk, 1 BS; Avon, 3 BS; Mt. Vernon, 2 BS; La Conner, 5 BS. *Island Co.*: San de Fuca, Whidby Island, 3 BS; Greenbank, Whidby Island, 2 BS; 3 mi. N Clinton, Whidby Island, 1 BS. *Snohomish Co.*: Oso, 2 BS; Hermosa Point, Tulalip Indian Reservation, 7 mi. W and 1/2 mi. N Marysville, 3 KU. *Chelan Co.*: Entiat, 2 BS. *Lincoln Co.*: 6 mi. E Odessa, 4 BS. *Spokane Co.*: Marshall, 7 BS. *Clallam Co.*: Neah Bay, 29 BS; 8 mi. W Sekin River, 1 WSC; mouth Sekin River, 1 WSC; Dungeness, 1 BS; Port Townsend, 3 BS; Ozette Indian Reservation, 1 CMNH; Sequim, 4 BS; Tivoli Island, Ozette Lake, 1 CMNH; Garden Island, Ozette Lake, 3 CMNH; Elwah, 1 WSC; Blyn, 1 BS; Soleduck River, 1 BS; 12 mi. S Port Angeles, 1 WSC; Forks, 9 CMNH, 1 BS; Cat Creek, 1 WSC; Lapush, 5 BS. *Jefferson Co.*: Jefferson Ranger Station, N Fork Hoh River, 5 CMNH; Duckabush, 6 BS. *Kitsap Co.*: Vashon Island, 2 BS. *King Co.*: Redmont, 2 BS; Kirkland, 20 BS; Seattle, 1 WSC, 3 KU; Northbend, 2 BS: Lake Washington, near Renton, 2 BS; Kent, 1 BS; Enumclaw, 1 BS. *Grays Harbor Co.*: Lake Quinault, 9 BS; Aberdeen, 20 BS; Westport, 5 BS, 2 WSC; Oakville, 1 BS. *Mason Co.*: Lake Cushman, 11 BS; Hoodsport, 1 BS;

North Fork Skokomish River, 1 BS; Shelton, 2 BS. *Pierce Co.*: Puyallup, 6 BS; Steilacoom, 1 BS; 6 mi. S Tacoma, 2 BS; Roy, 3 BS; Bear Prairie, Mt. Rainier, 1 BS; Reflection Lake, Mt. Rainier, 1 WSC. *Kittitas Co.*: Blewett Pass, 3 BS; Easton, 3 BS; 2 mi. E Cle Elum, 4 FC; Ellensburg, 2 BS. *Grant Co.*: Moses Lake, 1 BS; 9 mi. S and 1 mi. W Neppel, 1 UM. *Whitman Co.*: Hangman Creek, Tekoa, 1 WSC; 4 mi. ENE Pullman, 1 KU; 2 mi. N Pullman, 2 WSC; 2 mi. NW Pullman, 1 WSC; 2 mi. W Pullman, 1 WSC; Pullman, 5 WSC; Armstrong, 1 WSC; 5 mi. NE Wawawai, 1 BS; Wawawai, 5 WSC. *Thurston Co.*: Nisqually Flats, 2 BS; Nisqually, 1 BS; 4 mi. S Olympia, 1 BS; Tenino, 4 BS. *Pacific Co.*: Tokeland, 4 BS; 1 mi. S Nemah, 2 FC; 1 mi. N Bear River, Willapa Bay, 8 FC; 1/4 mi. N Bear River, 3 FC; 31/2 mi. E Seaview, 6 FC; Ilwaco, 1 BS. *Lewis Co.*: 8 mi. W Chehallis, 2 BS; Chehallis, 2 BS; Toledo, 1 BS. *Yakima Co.*: Selah, 7 KU; Wiley City, 4 BS. *Wahkiakum Co.*: Cathlamet, 1 BS. *Skamania Co.*: 45 mi. SE Toledo, 2 BS; Carson, 1 BS; Stevenson, 1 BS; 15 mi. NW White Salmon, 1 BS. *Klickitat Co.*: Trout Lake, 15 mi. S Mt. Adams, 2 BS; 15 mi. N Goldendale, 1 WSC; Goldendale, 1 BS. *Walla Walla Co.*: College Place, 1 KU. *Columbia Co.*: Starbuck, 3 BS. *Garfield Co.*: 1 mi. E Pomeroy, 1 SGJ. *Asotin Co.*: 21 mi. SE Dayton, 1 BS; Rogersburg, 1 BS.

IDAHO: *Bonner Co.*: 4 mi. S Sandpoint, 1 UM. *Kootenai Co.*: Coeur d'Alene, 2 BS. *Shoshone Co.*: Osburry, 1 BS; Mullan, 2 BS. *Latah Co.*: Felton's Mills, 1 WSC; Cedar Mtn., 5 WSC. *Lewis Co.*: Nezperce, 2 BS. *Idaho Co.*: Seven Devils Mtn., 1 BS. *Adams Co.*: Summit of Smith Mtn., 7500 ft., 5 KU; New Meadows, 1 BS; Tamarack, 1 BS. *Washington Co.*: 1 mi. NE Heath, SW slope Cuddy Mtn., 4000 ft., 7 KU. *Boise Co.*: Bald Mtn. R. S., 10 mi. S Idaho City, 1 BS. *Elmore Co.*: Cayuse Creek, 10 mi. N Featherville, 1 BS. *Canyon Co.*: Nampa, 5 BS. *Blaine Co.*: Sawtooth City, 5 BS; Alturas Lake, 1 BS. *Bonneville Co.*: 10 mi. SE Irwin, 5 BS. *Bannock Co.*: Pocatello, 1 BS, 1 KU; 1 mi. W Bancroft, 1 KU; Swan Lake, 1 BS. *Owyhee Co.*: Grasmere, 1 SGJ. *Cassia Co.*: 10 mi. S Albion, Mt. Harrison, 1 BS.

MONTANA: *Sanders Co.*: Prospect Creek, near Thomson Falls, 4 BS. *Lake Co.*: Flathead Lake, 5 BS. *Ravalli Co.*: Bass Creek, NW Stevensville, 2 BS; 2 mi. NE Stevensville, 1 UM; Corvallis, 4 BS; 6 mi. E Hamilton, 1 KU.

OREGON: *Clatsop Co.*: Seaside, 1 BS. *Washington Co.*: 5 mi. SE Hillsboro, 1 BS; Beaverton, 1 BS. *Multnomah Co.*: Portland, 20 BS; Portland, Switzler Lake, 5 BS. *Hood River Co.*: 2 mi. W Parkdale, 1 BS; north slope Mt. Hood, 2 BS. *Umatilla Co.*: 10 mi. W Meacham, 2 BS; Meacham, 3 BS. *Union Co.*: Elgin, 2 BS; Kamela, 2 BS; Hot Lake, 2 BS. *Wallowa Co.*: 25 mi. N. Enterprise, 4 BS; Wallowa Lake, 23 BS; S Wallowa Lake, 1 BS. *Clackamas Co.*: Estacada, 1 KU. *Marion Co.*: Salem, 8 BS; Permilia Lake, 2 BS. *Benton*

Co.: Corvallis, 2 BS; 5 mi. SW Philomath, 5 BS. *Linn Co.*: Shelburn, 1 BS. *Jefferson Co.*: 20 mi. W Warm Springs, 2 BS. *Grant Co.*: Beech Creek, 6 BS; Austin, 1 BS; Strawberry Butte, 1 BS; Strawberry Mts., 12 BS. *Baker Co.*: Homestead, 1 BS; Cornucopia, 11 BS; Rock Creek, 1 BS; Bourne, 7 BS; McEwen, 1 BS; Huntington, 1 BS; Anthony, 42 AMNH. *Lane Co.*: north slope Three Sisters, 3 BS; Vida, 1 BS; Mapleton, 1 BS; Eugene, 2 BS; 10 mi. S McKenzie Bridge, 1 BS; Florence, 1 BS. *Deschutes Co.*: Paulina Lake, 7 BS; Lapine, 8 BS. *Crook Co.*: 1 SGJ. *Douglas Co.*: Winchester Bay, 1 SGJ; Scottsburg, 3 BS; Drain, 5 BS; Lookingglass, 1 BS; Diamond Lake, 6 BS. *Coos Co.*: Empire, 5 BS. *Curry Co.*: Port Orford, 1 BS; Gold Beach, 4 BS. *Klamath Co.*: Anna Creek, Mt. Mazama, 1 BS; Crater Lake, 14 BS; Upper Klamath Marsh, 2 BS; Ft. Klamath, 35 BS; Klamath Falls, 6 BS. *Lake Co.*: 10 mi. SW Silver Lake, 3 BS; west fork Silver Creek, Yamsay Mts., 4 BS; Plush, 1 BS; Warner Creek, Warner Mts., 1 BS; Warner Mts., 3 BS; Gearhart Mts., 17 SGJ; *Harney Co.*: Diamond, 2 BS; Keiger Gorge, Steens Mts., 3 BS. *Malheur Co.*: 8 mi. W Jordon Valley, 1 BS.

WYOMING: *Lincoln Co.*: 13 mi. N and 2 mi. W Afton, 6 KU; 10 mi. N Afton, Salt River, 2 BS; 9 mi. N and 2 mi. W Afton, 6 KU; 7 mi. N and 1 mi. W Afton, 4 KU; Cokeville, 1 BS; 12 mi. N and 2 mi. E Sage, 2 KU; 6 mi. N and 2 mi. E Sage, 1 KU.

CALIFORNIA: *Del Norte Co.*: Smith River, 2 BS; Crescent City, 20 BS. *Siskiyou Co.*: Beswick, 1 BS; Hornbrook, 3 BS; Brownell, Klamath Lake, 1 BS; Mayten, 2 BS; Squaw Creek, Mt. Shasta, 5 BS; Upper Ash Creek, Mt. Shasta, 1 BS; upper Mud Creek, Mt. Shasta, 8 BS; Wagon Camp, Mt. Shasta, 5 BS; Warmcastle Soda Springs, Squaw Creek Valley, 2 BS; Sisson, 7 BS. *Modoc Co.*: Davis Creek, Goose Lake, 1 BS. *Humboldt Co.*: *Humboldt Bay*, 10 BS. *Trinity Co.*: Canyon Creek, 2 BS. *Shasta Co.*: Fort Crook, 11 BS; Dana, 17 BS; Fall Lake, Fall River Valley, 3 BS; Cassel, 2 BS; 12 mi. E Burney, 1 BS; Lassen Peak, 13 BS; Kellys, Warner Creek, 1 KU; Drakes Hot Springs, Warner Creek, 2 BS. *Mendocino Co.*: Russian Gulch State Park, 2 FC. *Plumas Co.*: 12 mi. NE Prattville, 2 BS; Spring Garden Ranch, Grizzly Mts., 3 BS; Sierra Valley, 1 BS. *Sierra Co.*: Lincoln Creek, 1 BS. *Sonoma Co.*: Petaluma, 3 BS; Point Reyes, 7 BS. *Placer Co.*: Donner, 3 BS. *El Dorado Co.*: Tallac, 3 BS. *Mono Co.*: Mt. Conness, 1 BS; Mono Lake, 1 BS; near Mammoth, 8000 ft., head of Owens River, 2 BS. *Inyo Co.*: Alvord, 1 BS.

NEVADA: *Elko Co.*: Mountain City, 1 BS; Three Lakes, 1 KU; west side Ruby Lake, 3 mi. N White Pine Co. line, 8 KU; Ruby Mts., 9 BS; W side Ruby Lake, 3 BS. *White Pine Co.*: W side Ruby Lake, 3 mi. S Elko Co. line, 1 KU. *Nye Co.*: Cloverdale, Reese River, 3 BS.

UTAH: *Weber Co.*: Beaver Creek, S Fork Ogden River, 2 UU; Huntsville, 10 mi. E Ogden, 1 UU; Hooper Bay Refuge, 4200 ft., 1 UU; Riverdale, 4200 ft., 3 UU; Riverdale, 4250 ft., 1 UU; 3 mi. SE Ogden, 2 UU; Snow Basin, 2 UU; Snow Basin, S part Wheeler Canyon, 1 UU; Uinta, 2 mi. W Weber Canyon entrance, 4 UU; 2 mi. W Uinta, 1 UU. *Salt Lake Co.*: City Creek Canyon, 6 mi. NE Salt Lake City, 4700 ft., 2 UU; 1 mi. up City Creek Canyon, 4600 ft., 1 UU; 3/4 mi. above Forks, City Creek Canyon, 1 UU; The Firs, Millcreek Canyon, 1 UU; Olympus Water Box, 1 UU; Salamander Lake, Lamb's Canyon, 9000 ft., 3 UU (near *obscurus*); Salt Lake City, 7500 ft., 1 UU; 1 mi. W Draper, 4500 ft., 6 UU; Draper, 4500 ft., 5 UU; 11/2 mi. SW Draper, 4500 ft., 1 UU; 3 mi. SW Draper, 4400 ft., 2 UU; 3 mi. S Draper, 4400 ft., 2 UU; 1 mi. S Draper, 4500 ft., 1 UU. *Juab Co.*: W side Deep Creek Mts., Queen of Sheba Canyon, 8000 ft., 3 UU. *Wasatch Co.*: Midway Fish Hatchery, 5450 ft., 1 UU.

Marginal records.--BRITISH COLUMBIA: Okanagan; Westbridge; Kuskonook; Cranbrook. MONTANA: Flathead Lake; 6 mi. E Hamilton; Prospect Creek. IDAHO: Cedar Mtn.; New Meadows; Alturas Lake; 10 mi. SE Irwin. WYOMING: 13 mi. N and 2 mi. W Afton; 6 mi. N and 2 mi. E Sage. IDAHO: 1 mi. W Bancroft; Swan Lake. UTAH: Beaver Creek, South Fork, Ogden River; Midway Fish Hatchery; west side Deep Creek Mts., Queen of Sheba Canyon, 8000 ft. NEVADA: Baker Creek (Hall, 1946:120); Reese River (*ibid.*); 2 mi. S Hinds Hot Springs (*ibid.*). CALIFORNIA: Mono Lake (Jackson, 1928:110); near Mammoth; Alvord; Mount Conness; Donner; Buck Ranch (Jackson, 1928:110); Warner Creek, Drake Hot Springs (*ibid.*); Canyon Creek; Cuddeback (Jackson, 1928:105); Novato Point (*ibid.*), thence northward along the coast to WASHINGTON: Friday Harbor, San Juan Island. BRITISH COLUMBIA: Port Moody.

Sorex vagrans obscuroides new subspecies

Type.--First year female, skin and skull; No. 30064/42074, U. S. Biol. Surv. Coll.; obtained on August 9, 1891, by Frank Stephens from Bishop Creek, 6600 ft., Inyo Co., California, original no. 811.

Range.--The Sierra Nevada of California, north at least to El Dorado County, intergrading northerly with *S. v. vagrans*.

Diagnosis.--Size medium for the species; average and extreme measurements of 5 topotypes are: total length, 107 (103-112); tail, 47 (45-50); hind foot, 12.8 (12-13.5). Skull relatively broad interorbitally; color of dorsum in summer pelage nearest (17 *k*) Olive Brown.

Comparisons.--Differs from *S. v. vagrans,* with which it intergrades to the north, in: longer tail and total length; skull larger and relatively broader interorbitally; color in summer grayer (less reddish), the lighter subterminal color bands of the hair often showing through the darker tips and imparting a grizzled appearance to the dorsum. Differs from *S. v. parvidens* to the south in: skull relatively broader interorbitally and less flattened; teeth slightly larger.

Remarks.--*S. v. obscuroides* has long been called *S. v. obscurus*. In fact, *obscuroides* is separated from the range of *obscurus* by the intervening, smaller subspecies *S. v. vagrans. S. v. obscuroides* resembles *S. v. obscurus* in color and size but the skull is smaller, although relatively slightly broader. The resemblance in color is possibly due to the fact that *obscuroides,* like *obscurus,* is a high mountain form. *S. v. obscuroides* intergrades with *S. v. vagrans* along the crest of the Sierra between Yosemite National Park and Lassen Peak and on the eastern slope of the Sierra from approximately Mammoth northward. Specimens from Donner are intergrades but are closest to *S. v. vagrans*. Although all specimens from Lassen Peak are referable to *S. v. vagrans,* some show cranial characters of *obscuroides*.

Specimens examined.--Total number, 76. CALIFORNIA: *Mono Co.*: Mt. Dana, 6 BS; Mt. Lyell, 11 BS. *Mariposa Co.*: Tuolumne Meadows, Muir Meadow, 9300 ft., 1 BS; Tuolumne Meadows, Mt. Unicorn, 1 BS; Tuolumne Meadows, N base Mt. Lyell, 8 BS; Tuolumne Meadows, Soda Springs, 4 BS; Lake Tenaya, 5 BS. *Madera Co.*: San Joaquin River, 8000 ft., 4 BS. *Fresno Co.*: Horse Corral Meadows, 3 BS. *Mono Co.*: head of Owens River near Mammoth, 2 BS. *Inyo Co.*: Bishop Creek, 5 BS; Round Valley, 1 BS. *Tulare Co.*: E. Fork Kaweah River, 7 BS; Mt. Whitney, 5 BS; Whitney Creek, Mt. Whitney, 4 ChM; Whitney Meadows, 9700 ft., 1 BS; Mineralking, 2 BS; N. Fork Kern River, 9600 ft., 1 BS; S. Fork Kern River, 4 BS; Kern Lakes, 1 BS.

Marginal records.--CALIFORNIA: Pyramid Peak; near Mammoth; *Round Valley*; Bishop Creek; Mt. Whitney; Kern Lakes; Halstead Meadows; Horse Corral Meadows; east fork Indian Canyon (Jackson, 1928:121).

Sorex vagrans parvidens Jackson

Sorex obscurus parvidens Jackson, Jour. Mamm., 2:161, August 19, 1921.

Type.--Adult male, skin and skull; No. 56561, U. S. Biol. Surv. Coll.; obtained on October 3, 1893, by J. E. McLellan from Thurmans Camp, Bluff Lake, 7500 ft., San Bernardino Mts., California.

Range.--Confined, so far as known, to the San Bernardino and San Gabriel mountains, San Bernardino Co., California.

Diagnosis.--Size medium for the species; measurements of two specimens from the San Bernardino Mountains are: total length, 105, 106; tail, 41, 48; hind foot, 12, 14. Upper parts in summer Olive-Brown to Buffy-Brown; cranium flattened and relatively narrow; unicuspids and incisors relatively small.

Comparisons.--For comparison with *S. v. obscuroides,* the only adjacent subspecies, see the account of that subspecies.

Remarks.--S. v. parvidens is seemingly an uncommon mammal. I have been informed by Terry Vaughan that repeated attempts by him to obtain it in suitable habitat in the San Gabriel Mountains failed. This shrew is probably no longer in reproductive continuity with *Sorex vagrans* of the Sierra Nevada.

Specimens examined.--Total number, 4. CALIFORNIA: *San Bernardino Co.*: type locality, 4 BS.

Marginal records.--CALIFORNIA: Camp Baldy, San Antonio Canyon (Jackson, 1928:124); type locality.

Sorex vagrans halicoetes Grinnell

Sorex halicoetes Grinnell, Univ. California Publ. Zool., 10:183, March 20, 1913.

Sorex vagrans halicoetes, Jackson, N. Amer. Fauna, 51:108, July 24, 1928.

Type.--Young adult male, skin and skull; No. 3638, Mus. Vert. Zool.; obtained on May 6, 1908, by Joseph Dixon from salt marsh near Palo Alto, Santa Clara Co., California.

Range.--Marshes in the southern part of San Francisco Bay, California.

Diagnosis.--Size small for the species; measurements of two topotypes are: total length, 105, 106; tail, 39, 40; hind foot, 12, 13. Upper parts in winter Chaetura Black or near Fuscous-Black; underparts brownish; upper parts in summer near (17 *m*) Mummy Brown; underparts with a decided buffy wash, near (15 *d*) Light Ochraceous Buff; rostrum relatively large; maxillary tooth-row relatively long; teeth relatively large.

Comparisons.--Darker ventrally, both summer and winter, than *S. v. vagrans;* slightly more reddish dorsally in summer pelage than *S. v. vagrans,* rostrum and teeth relatively larger; smaller externally than *S. v. paludivagus,* paler; skull longer, narrower cranially and broader rostrally.

Remarks.--This subspecies seems to be restricted to salt marshes where it occurs with *Sorex ornatus*.

Specimens examined.--Total number, 12. CALIFORNIA: *San Francisco Co.*: San Francisco, 4 BS. *Alameda Co.*: West Berkeley, 1 BS; Berkeley, 1 BS; Dumbarton Point, 1 KU. *San Mateo Co.*: San Mateo, 2 BS. *Santa Clara Co.*: Palo Alto, 3 BS.

Marginal records.--CALIFORNIA: Berkley, *Elmhurst*; *Palo Alto*; San Mateo.

Sorex vagrans paludivagus von Bloeker

Sorex vagrans paludivagus von Bloeker, Proc. Biol. Soc. Washington, 52:93, June 5, 1939.

Type.--Adult male, skin and skull; No. 5053, Los Angeles Museum of History, Science and Art, obtained on November 3, 1938, by Jack C. von Bloeker, Jr., from salt marsh at mouth of Elkhorn Slough, Moss Landing, Monterey Co., California, original no. 9456.

Diagnosis.--Size medium for the species; average and extreme measurements of 6 topotypes are: total length, 115 (113-118); tail, 46.5 (42-48); hind foot, 14.5 (14-15) (von Bloeker, 1939:94). In winter nearly black dorsally, deep mouse gray ventrally; in summer nearly as dark dorsally as in winter, hairs of venter tipped with Clove Brown; skull short, relatively broad cranially and relatively narrow rostrally.

Comparisons.--For comparison with *S. v. halicoetes* see account of that subspecies.

Remarks.--This subspecies, occurring at the limits of the range of the species, is uncommon in most collections. Seven specimens were available for the original description. The summer pelage is not completely described in the original description, but is stated to be darker than the winter pelage of *S. v. vagrans*, and must thus be considerably darker than the summer pelage of *S. v. halicoetes*. Two specimens in the Museum of Vertebrate Zoology, from San Gregario, referred by Jackson to *S. v. halicoetes*, were included in the present subspecies by von Bloeker.

Specimens examined.--None.

Records of occurrence (von Bloeker, 1939:94).--CALIFORNIA: *San Mateo Co.*: San Gregario. *Monterey Co.*: Seaside; mouth of Salinas River; Moss Landing.

Marginal records.--CALIFORNIA: San Gregario; Seaside.

Sorex vagrans vancouverensis Merriam

Sorex vancouverensis Merriam, N. Amer. Fauna, 10:70, December 31, 1895.

Sorex vagrans vancouverensis, Jackson, N. Amer. Fauna, 51:106, July, 1928.

Type.--Adult male, skin and skull; No. 71913, U. S. Biol. Surv. Coll.; obtained on May 10, 1895, by Clark P. Streator, from Goldstream, Vancouver Island, British Columbia.

Range.--Vancouver Island from Sayward south, and Bowen Island.

Diagnosis.--Size small for the species; average and extreme measurements of 6 specimens from Alberni Valley, Vancouver Island, are: total length, 106.5 (97-115); tail, 41.7 (40-43); hind foot, 12 (11-13) (Jackson, 1928:107). Ventral parts brownish, winter pelage reddish brown rather than grayish.

Comparisons.--Differs from *S. v. vagrans* in more brownish ventral parts and more brownish, rather than grayish, winter pelage; differs from the sympatric *S. v. isolatus* in shorter tail, shorter hind foot, more narrow skull, and smaller teeth.

Remarks.--This is a poorly differentiated subspecies which is closely related to *S. v. vagrans*. The differences in color noted are average ones. Some individuals of this shrew might be difficult to separate from *S. v. isolatus*. The slight degree of morphological divergence is such that intergrades might be expected to occur. Possibly some habitat separation occurs, but such has not been reported.

Specimens examined.--Total number, 3. BRITISH COLUMBIA: Vancouver Island: Mt. Washington, 1 KU; Nanaimo, 1 BS; type locality, 1 BS.

Marginal records.--BRITISH COLUMBIA: Sayward (Anderson, 1947:18); Bowen Island (Hall, 1938:463); Alberni (Jackson, 1928:107).

CONCLUSIONS

1.

Sorex vagrans, S. obscurus, S. pacificus, and *S. yaquinae* are conspecific with one another. Each is a valid subspecies but all should bear the specific name *Sorex vagrans* Baird, 1858.

2.

The subspecies of *Sorex vagrans* form a cline from large (*pacificus*) to small (*vagrans*). The cline is bent in such a manner that the terminal subspecies occur together. Where the two subspecies occur together, individuals of one subspecies do not crossbreed with individuals of the other subspecies and therefore react toward one another as do full species. *Sorex vagrans vagrans* occurs sympatrically with *S. v. sonomae, S. v. pacificus, S. v. yaquinae, S. v. bairdi, S. v. permiliensis,* and *S. v. setosus. S. v. vancouverensis* occurs sympatrically with *S. v. isolatus.*

3.

The sympatric existence of the terminal subspecies of the *Sorex vagrans* rassenkreis is made possible by marked differences between them in size and in ecological preference.

4.

The west-coast subspecies, *sonomae, pacificus, yaquinae, bairdi,* and *permiliensis* probably differentiated from the Great Basin and Rocky Mountain subspecies, *vagrans, obscurus* and *monticola,* during a separation caused first by aridity in the Great Basin, and secondly by glaciation of the Cascade Mountains and the Sierra Nevada, possibly in the Sangamonian and Wisconsinan ages respectively.

5.

Sorex v. vagrans originated in the Great Basin and arrived on the Pacific Coast after the last deglaciation of the Cascades and Sierra Nevada.

6.

In *S. vagrans,* heterogonic growth is illustrated; the larger the skull, the larger the rostrum in proportion to the skull as a whole.

7.

In the species *S. vagrans,* size and color vary geographically more than do other features.

8.

The *S. ornatus* group, *S. longirostris,* and *S. veraepacis* had a common ancestor with *S. vagrans,* possibly in the Illinoian Age.

9.

S. vagrans, the *S. ornatus* group, *S. veraepacis, S. longirostris, S. palustris, S. bendiri,* and the *S. cinereus* group, because of structural resemblances, should be placed in a single subgenus, *Otisorex. S. trowbridgii,* the *S. arcticus* group, the *S. saussurei* group, *S. merriami, S. fumeus,* and *S. dispar,* should be included in the subgenus *Sorex.*

10.

Sorex cinereus occurs with the medium-sized and large-sized *S. vagrans* in the Rocky Mountains and in Canada, but does not occur with the smaller subspecies of *S. vagrans,* probably because competition between two shrews of like size excludes *S. cinereus.*

TABLE 1--CRANIAL MEASUREMENTS OF SOREX VAGRANS

Catalog number or number of individuals averaged	Condylobasal length	Palatal length	Maxillary tooth-row	Cranial breadth	Least interorbital breadth	Maxillary breadth
Sorex vagrans pacificus, Orick, California.						
8 av	21.8	9.6	8.6			

10.4
4.1
6.6

Max

22.8
10.2
9.0
11.1
4.3
6.8

Min

21.3
9.3
8.4
10.2
4.1
6.4

Sorex vagrans yaquinae, Newport, Oregon.
707 AW

20.1
8.9
7.6
9.3
3.7
5.7

706 AW

19.3
8.8
7.3
9.3
4.0
5.8

Mapleton, Oregon.
205273 USBS

20.6
9.0
8.2

9.9

4.2

6.0

205270 USBS

20.4

8.9

7.9

9.3

3.7

6.0

205272 USBS

9.2

8.3

4.0

6.1

Vida, Oregon.

4 av

19.5

8.4

7.5

9.2

3.6

5.5

Max

20.3

8.8

7.9

9.5

3.7

5.7

Min

19.3

8.2

7.3

8.9

3.5

	5.3

McKenzie Bridge, Oregon.	
6 av	
	18.9
	8.2
	7.1
	9.1(5)
	3.7
	5.6
Max	
	19.5
	8.6
	7.6
	9.4
	3.8
	5.7
Min	
	18.7
	8.0
	6.8
	8.4
	3.6
	5.4

Sorex vagrans bairdi, Astoria, Oregon.	
6 av	
	18.5
	7.8
	7.0
	8.9(4)
	3.4
	5.3
Max	
	19.2
	8.1
	7.4
	9.0
	3.5

5.5

Min

18.0
7.6
6.9
8.9
3.2
5.2

Sorex vagrans permiliensis, Mt. Jefferson, Oregon.
14 av

18.0
7.5
6.8
9.0
3.5
5.2

Max

18.9
7.9
7.2
9.4
3.7
5.3

Min

17.2
7.2
6.5
8.6
3.3
4.8

Sorex vagrans setosus, Olympic Mts., Washington.
12 av

17.2(9)
7.1
6.5
8.5(8)
3.3

	5.0
Max	
	17.9
	7.4
	6.8
	8.7
	3.4
	5.3
Min	
	16.7
	6.9
	6.2
	8.3
	3.0
	4.8

Mt. Rainier, Washington.	
16 av	
	17.2
	7.1
	6.5
	8.4(14)
	3.3
	5.0
Max	
	17.6
	7.3
	6.7
	8.7
	3.5
	5.2
Min	
	16.4
	6.5
	6.1
	8.1
	3.2
	4.7

Sorex vagrans longicauda, head Rivers Inlet, B.C.

15 av

18.0
7.4
6.7
8.7
3.2
5.0

Max

18.4
7.6
6.9
8.9
3.3
5.2

Min

17.6
7.2
6.4
8.4
3.1
4.8

Port Simpson, British Columbia.

10 av

18.1(9)
7.6
7.0
8.9
3.4
5.1

Max

18.8
7.8
7.2
9.2
3.6
5.4

Min

17.2
7.2
6.6
8.5
3.3
4.9

Fort Wrangell, Alaska.
18 av

18.5
7.8(15)
7.1
9.0(15)
3.3
5.1

Max

18.9
8.0
7.3
9.2
3.5
5.3

Min

17.8
7.5
6.7
8.6
3.2
5.0

Sorex vagrans elassodon, Woewodsky Is., Alaska.
20550 AMNH

18.0
7.5
6.7
8.7
3.3
5.0

20553 AMNH

17.5
7.1
6.3
8.3
3.2
4.7

Sorex vagrans alascensis, 9 mi. W and 4 mi. N Haines, Alaska.
10 av

17.2
7.2
6.7
8.5
3.1
4.9

Max

17.6
7.4
6.9
8.8
3.3
5.0

Min

16.9
6.9
6.5
8.2
3.0
4.7

Yakutat Bay, Alaska.
73543 USBS

18.0
7.5
6.8
...
3.2
5.0

73536 USBS

18.0
7.6
6.8
8.8
3.4
5.3

73541 USBS

17.9
7.4
6.7
8.8
3.1
5.2

Sorex vagrans shumaginensis, Sandpoint, Popof Is., Alaska.
9 av

17.2(5)
7.0
6.3
8.3(7)
3.1
4.8

Max

17.6
7.2
6.6
8.5
3.2
5.1

Min

16.8
6.8
6.1
8.0
3.0
4.7

Sorex vagrans obscurus, Barkerville, British Columbia.
5 av

17.1
7.1
6.5
8.5(4)
3.3
4.8

Max

17.3
7.3
6.6
8.6
3.3
5.0

Min

16.6
6.7
6.4
8.2
3.2
4.6

10 mi. SSW Leadore, Idaho.
7 av

17.2(4)
7.3(9)
6.6
8.6(4)
3.3
5.0

Max

17.3
7.5
6.8
8.9
3.4
5.1

Min

17.0
7.1

6.4
8.3
3.2
4.7

Albany Co., Wyoming (several localities).
20 av

17.3
7.3
6.8
8.7(19)
3.2
5.2

Max

17.9
7.6
6.9
9.0
3.4
5.5

Min

16.7
6.9
6.5
8.4
3.1
5.0

Sorex vagrans longiquus, 25 mi. ESE Big Sandy, Montana.
87332 UM

16.4
6.8
6.2
8.2
3.0
4.8

87334 UM

16.8
7.1

	6.3
	8.1
	3.2
	4.7
87335 UM	
	15.8
	6.7
	6.0
	8.4
	3.1
	4.9

Highwood Mts., Montana.	
10 av	
	16.3(9)
	6.7
	6.2
	8.0(9)
	3.1
	4.7
Max	
	16.9
	6.9
	6.4
	8.3
	3.3
	5.0
Min	
	15.6
	6.5
	6.0
	7.8
	3.0
	4.5

Sorex vagrans neomexicanus, Cloudcroft, New Mexico.	
4 av	
	17.6(3)
	7.6

7.0
8.7
3.3
5.2

Max

17.7
7.7
7.1
8.8
3.4
5.4

Min

17.4
7.4
7.0
8.5
3.2
5.1

Sorex vagrans monticola, White Mts., Arizona.

12 av

16.1(9)
6.6
5.9(9)
8.2
3.1(11)
4.7

Max

16.6
7.0
6.1
8.5
3.3
4.9

Min

15.5
6.5
5.6
8.1

3.0
4.6

Sorex vagrans orizabae, Volcan Toluca, Mexico.

55900 USBS

17.1
6.9
6.2
7.8
2.9
4.5

55898 USBS

17.1
6.8
6.1
8.0
3.0
4.8

55897 USBS

16.8
6.9
6.1
7.9
2.9
4.6

Sorex vagrans vagrans, Lincoln Co., Wyoming.

7 av

16.5(6)
6.6
6.1
8.2
2.9
4.7

Max

17.1
7.0
6.4
8.5

3.1
4.9

Min

16.0
6.4
5.9
7.9
2.9
4.5

Gearhart Mtn., Lake Co., Oregon.
17 av

16.5(15)
6.6
5.9
8.1
2.9
4.6

Max

17.1
7.0
6.6
8.5
3.1
4.9

Min

16.1
6.2
5.7
7.8
2.8
4.4

Willapa Bay, Washington.
9 av

16.6
6.8
6.2
8.1

2.8
4.7

Max

17.2
7.1
6.4
8.3
3.1
5.0

Min

16.2
6.6
5.9
7.9
2.7
4.6

Sorex vagrans obscuroides, Bishop Creek, California.

4 av

16.7
7.0
6.3
8.2
3.2
4.8

Max

16.8
7.1
6.4
8.3
3.4
4.9

Min

16.6
6.9
6.2
8.1
3.1
4.7

Mt. Whitney, California.
4 av

16.7(3)
6.9
6.3
8.4
3.3
4.8

Max

16.7
7.0
6.4
8.5
3.4
5.0

Min

16.7
6.7
6.1
8.4
3.1
4.7

Sorex vagrans parvidens, San Bernardino Peak, California.
56559 USBS

17.1
7.0
6.1
8.0
2.9
4.8

56558 USBS

16.4
6.8
6.1
8.0
3.0

4.8

Made in the USA
Monee, IL
08 November 2026

68030007R00057